Baja California's Coastal Landscapes Revealed

Other Books by Markes E. Johnson

Discovering the Geology of Baja California: Six Hikes on the Southern Gulf Coast (2002)

Off-Trail Adventures in Baja California: Exploring Landscapes and Geology on Gulf Shores and Islands (2014)

Co-editor with Jorge Ledesma-Vázquez

Atlas of Coastal Ecosystems in the Western Gulf of California (2009)

Available from the University of Arizona Press

Gulf of California Coastal Ecology: Insights from the Present and Patterns from the Past

Available from Sunbelt Publications

Baja California's Coastal Landscapes Revealed

Excursions in Geologic Time and Climate Change

Markes E. Johnson

THE UNIVERSITY OF
ARIZONA PRESS
TUCSON

For science teachers, everywhere and on all levels, but especially in memory of Clifford O. Johnson (1910–1991), consummate high school biology teacher, summer sailor with the Naval Reserve following wartime service in the U.S. Navy, and patient father.

The University of Arizona Press
www.uapress.arizona.edu

We respectfully acknowledge the University of Arizona is on the land and territories of Indigenous peoples. Today, Arizona is home to twenty-two federally recognized tribes, with Tucson being home to the O'odham and the Yaqui. Committed to diversity and inclusion, the University strives to build sustainable relationships with sovereign Native Nations and Indigenous communities through education offerings, partnerships, and community service.

ISBN-13: 978-0-8165-4252-9 (paperback)

Cover design by Leigh McDonald
Cover image by Markes E. Johnson

Unless otherwise noted, all photos are by the author.

Library of Congress Cataloging-in-Publication Data
Names: Johnson, Markes E., author.
Title: Baja California's coastal landscapes revealed : excursions in geologic time and climate change / Markes E. Johnson.
Description: Tucson : University of Arizona Press, 2021. | Includes bibliographical references and index.
Identifiers: LCCN 2021012043 | ISBN 9780816542529 (paperback)
Subjects: LCSH: Geology—Mexico—Baja California (Peninsula)—Guidebooks. | Climatic changes—Mexico—Baja California (Peninsula)—Guidebooks. | Baja California (Mexico : Peninsula)—Guidebooks. | LCGFT: Guidebooks.
Classification: LCC QE203.B34 J637 2021 | DDC 557.2/2—dc23
LC record available at https://lccn.loc.gov/2021012043

Printed in the United States of America
♾ This paper meets the requirements of ANSI/NISO Z39.48-1992 (Permanence of Paper).

Contents

Illustrations

Figures

Maps

Plates

Preface

On the Importance of Rock Reading

HAVING TAUGHT A range of introductory and advanced geology courses through an active career at a leading liberal-arts college, I am proud of those former students who became teachers plying the same trade at various levels in secondary schools, colleges, and major universities. Most of my students, however, took up careers with little or no connection to the earth sciences. I always began my first lecture on historical geology with a message of inclusivity. Anyone can learn how to read the rocks. I promised that once a fluency in rock reading was mastered, students could count on a life skill allowing them to better appreciate their physical surroundings, wherever they might choose to live or travel in future years. When a very few complained about having to memorize the vocabulary relevant to rock reading, I reminded them how foolish it might sound to bemoan the workload in a Spanish or German course that required a mastery of words as the fundamental building blocks for sentences in those languages. I argued that you can't study the grammar of a language without taking the time to learn the words enfolded within that grammar.

The best part of being a professor is that those who teach never really stop being a student. That is to say, every day brings fresh opportunities to learn something new that might be added to a font of knowledge and experience. Naturally, it is most rewarding to become a lifelong student while also earning a salary for doing so. The learning process need not stop with retirement from teaching duties. I distinctly recall from my undergraduate days as a geology student at the University of Iowa in Iowa City, just how impressed I was by two, white-haired, emeriti professors who showed up most week days to climb three flights of stairs to attic cubby-holes in Calvin Hall, where they carried on with their studies. Despite the fact that their teaching days were over, their names were well known to us. Whenever I spied them climbing the stairs early in the day, I rejoiced in the thought that I might join a profession holding out a promise to keep my muscles in tone and my mind alert well into my senior years. I have now earned the *gravitas* of a white tonsure (and white beard to accompany it), and I have the luxury of looking back on a career that witnessed several changes in direction, affording an expanded outlook on the world.

Even the most rudimentary attempt to read the rock record confronts two difficulties. The first is the fragmentary story conveyed by any single pile of layered rocks exposed at any one place around the world, and the second is the shear immensity of geologic time. From *On the Origin of Species*, 1859, p. 310, Charles Darwin likened rock layers to a book: "Of this volume, only here and there a short chapter has been preserved; and of each page, only here and there a few lines." In a later edition of his famous treatise, Darwin realized that the second problem required better explication, and he repeated sound advice on how to put the enormity of geologic time into perspective:

> Take a narrow strip of paper, 83 feet 4 inches in length, and stretch it along the wall of a large hall; then mark off at one end the tenth of an inch. This tenth of an inch will represent one hundred years, and the entire strip a million years.

I, myself, adapted this example to the classroom, after spending a weekend cutting out and gluing together strips of unused wallpaper to make a continuous piece amounting to the prescribed length. When installed, the display took up three sides of the lecture hall in Clark Hall at Williams College, where my introductory course in historical geology was offered. To represent the roughly five-million-year record of sedimentary rocks that accumulated around the Gulf of California after seawater flooded into the growing rift between the Mexican mainland and Baja California peninsula, such a strip of paper would need to wrap around the walls of lecture hall at least three and a half times.

The comparison with a fragmentary book is apt, but the challenge may be overcome by finding the most complete rock sequence that a particular region has to offer. Such a process of exploration might be likened to visits to any number of bookstores in a dedicated search for a rare book, or even an effort to piece together all the additions and changes though the several editions of a book like Darwin's *Origin of Species*. After 30 years of exploration throughout the Baja California peninsula, it is quite possible that the most complete succession of Pliocene and Pleistocene strata was discovered by my research group on Isla del Carmen (see chapter 5, this volume). By no means was this search based on my intuition alone, but benefited from generations of geologists who preceded us.

The sequence at Arroyo Blanco on the east coast of Isla del Carmen amounts to strata 200 ft (61 m) in thickness. Even here, however, there occur breaks or pauses in the succession that point to missing pages, if not short chapters in the local story. Yet another consideration is the style of sedimentary accumulation that any particular set of strata was subject to. In short, some of the most complete pages in the rock record entail what is called ecologic time. Preservation of a former oyster bed or coral reef in intricate detail that records the succession of one generation after another are good examples of ecologic time. It all amounts to the same story in today's world that might be observed by repeated visits to an oyster bank or coral reef over a stretch of 50 years of uninterrupted growth. I confess that during my earlier visits to the Gulf of California, discovery of sites where ecologic time was found to be preserved in the rock record met with the same excitement reserved for recovering the Holy Grail. The effect was like recovering a few frames of film in a nature movie, each frame offering a glimpse into a single moment in time captured forever from a remote past.

More commonly, limestone follows a pattern whereby accumulations accrue by bits and pieces of oysters or corals fragmented after death to make a coquina or a kind of organic conglomerate with few complete fossils. The effect is more like an amalgamated set of photographic frames distorted and pressed together to record a blurry image. Some outside perturbation of normal, everyday life took place to disrupt the flow of time and skew the film. In part, this is what Darwin was getting at with his insistence that the rock record is imperfect. He believed that organic evolution typically unfolded in a gradual manner, and the motif of the imperfect book suited his frustration with fossil lineages that often failed to show all the intermediate stages of change with time.

Inorganic rocks, like shale, sandstone, and conglomerate, reflect the passage of a parallel cycle, whereby larger chunks of rock are ground down in size by abrasion one against another to produce smaller bits. Darwin, too, offered practical insights as to how such a process occurred in the foment of geologic time. He argued that a person might revisit a coastline on a regular basis and witness very little change, day-by-day, through the passage of untold tidal cycles. But then, notable changes might be detected after the landfall of a major sea storm. One might visit the same shore, season-after-season and year-after-year, observing that only modest change resulted from "normal" storms. The storm of a century, however, is usually outside the experience of a single person. Yet repetition of such storms over a thousand years or over a million years may inflict substantial degradation along a given stretch of shore, as the reduction of boulders and cobbles into sand occurs in fits and spurts over time.

The Gulf of California (also known as the Sea of Cortez) is a deep body of water between the Mexican mainland and the Baja California peninsula, with a northwest-southeast axis that extends for 685 miles (1,100 km) from the Colorado River delta in the north to a 112-mile (180-km) wide opening to the Pacific Ocean in the south. It is a place of largely untamed wildness and great beauty. It also exists as an extraordinary laboratory posited on a colossal scale that affords all kinds of studies and comparisons at the intersection of geography and ecology with the paleogeography and paleoecology of the past. While my career has taken me to many other places around the world, it is to Baja California that I return again and again each year.

This is the third installment in a trilogy offered by the University of Arizona Press, all three of which are intended for the naturalist who recognizes the overlap of the present and the past in a setting yet as pristine as the Gulf of California. The first installment (*Discovering the Geology of Baja California*, 2002) focused on a single area centered around Punta Chivato that juts out into the Gulf of California as a rocky promontory. A series of six hikes, some more demanding than others, provides commentary on the local geology and paleontology. I argued that the landscape around Punta Chivato is a microcosm representative of the entire Gulf of California. Indeed, the physical processes that shape the promontory today and did so in the past conform to a set of physical attributes dominated by seasonal wind patterns and wave surge to which intertidal life has responded the same way throughout much of the gulf region. Punta Chivato is endowed with the kind of limestone deposits that capture ecologic time in well-preserved fossil deposits.

The second book in the sequence (*Off-Trail Adventures in Baja California*, 2014), followed much the same tactic, with eight hikes, each of which focused on a different area spread out along the gulf from north to south. The aim was to make more implicit the controlling physical factors that govern the gulf's coastal life now and in the geologic past, while still celebrating the extraordinary circumstances that led to the preservation of fossil assemblages frozen in ecologic time. However, the final excursion in that book took a hard look at the impact of hurricanes in the southern Gulf of California, specifically Isla Cerralvo, located east of La Paz. There, fossil coral reefs still reflect ecologic time, but the larger story was one of a landscape torn inside out by the violence of intense storms that arrived on an episodic schedule. Our research team was effectively sidetracked by the lasting signs of such storms, some of which could be pinned to historic hurricanes that ravaged the island as recently as 2006. After several visits to the island, the overall experience shaped expectations of finding similar patterns in the sediments and sedimentary rocks found to the north. Other places that had been puzzling, suddenly made sense in the context of Isla Cerralvo.

Hence, the outlook of this third volume is decidedly different, being focused on the impact of the rare and violent event (rare at least in terms of a human life span), if not catastrophic in character. The model remains the same, organized into chapters that describe different areas through the vehicle of a guided excursion, but spread out from one end of the Gulf of California to the other. My hope is that the reader will become more proficient in the vocabulary of big storm events and the kinds of rock formations that result from their impact. Some of that vocabulary resides in the public consciousness. Who among us is not familiar with the grading of big storms by intensity as a hurricane 1 to 5? That such a scheme belongs to the Saffir-Simpson Hurricane Wind Scale may be new, but the background already exists in the popular imagination.

The importance of rock reading in this context is vital to everyone on Planet Earth. Global warming is a phenomenon much in the news. The events described from the past in this telling occurred well before human beings had the capability to alter climate. Indeed, it will be shown that such events from the Pliocene and Pleistocene are characteristic of intervals when the global climate was more extreme than today. One can get mired in the debate over whether we humans are responsible for the increase in global temperature and rising sea levels, but the fact remains that the present rate of change through only the last few decades exceeds anything recorded in the recent geologic past. By gauging the ferocity of major storms from the past, we may at least be warned about the potential for changes in climate that threaten us all during the rest of this century. It is said that the present is a key to the geologic past, which usually means that the same laws of physics, laws of chemistry, and laws of biology that we see around us today operated the same way through most of geologic time stretching back at least 500 million years into the past. But the inverse relationship also gives pause for thought. The past, it seems, may be a reliable guide to the future.

I invite you to join me on a series of rambles through some of the most unusual landscapes of Baja California that I have thought deeply about since publication of the previous books with annotated hikes. These I call rambles, because they are shorter hikes. Although the body remains reasonably fit and the mind as keen as ever, I am no longer game for a 10.5-mile (17-km) hike over rough country at a single go. Of course, the armchair participant is welcome to come along. For those who would follow in my footsteps, I offer the same advice as before. Always hike with a companion who can go for help in an emergency. Always look where you place each foot in crossing uneven terrain. Never climb through rocks placing a hand above your head out of eyesight, where a creature capable of a debilitating bite might be surprised. Be sure to plan outings with sufficient water and snacks to get you through the day. Always inform a third party where you plan to hike and when you expect to return at the close of day. These are common-sense precautions that serve the explorer well. Nearly eighty Williams College students have accompanied me on hikes around Baja California. More than that number of college alumni have done so, as well. Four of the hikes described herein were field-tested by willing subjects. All returned home unscathed with stories to tell.

The vocabulary needed to read the stories told by rocks and fossils can be a challenge, but the glossary of words at the back of this book should make things clear where my dialog is less than definitive.

Markes E. Johnson

Williamstown, MA

July 31, 2020

Acknowledgments

The background momentum leading to this volume benefited from the support of officials at Williams College who facilitated a postretirement return to the teaching faculty for the spring 2016 semester to offer a course on the geology and marine ecology around the Gulf of California. My recruitment included supplemental funding for a two-week student field trip to the Loreto area. Without exception, the dozen students who participated in that course were among the brightest and most enthusiastic I have had the pleasure to instruct. Their final mapping project at San Basilio had a decisive impact on the outcome of chapter 4 in this book. Additional travel support from the Science Division at Williams College for a separate trip to Mexico later that spring for fieldwork in the Puertecitos area in the upper Gulf of California led to the research described in chapter 2. I am especially grateful for the use of a campus office during the planning, map drafting, and writing stages of this book. Moreover, the Office of the Dean of Faculty at Williams College contributed a generous subsidy to the University of Arizona Press that helped make this book possible.

Many individuals contributed privately to help bring this project to fruition. Chief among them are Norman K. Christie and his spouse Maxine, part-time residents of Loreto in Baja California Sur, who provided housing for students during our spring 2016 visit and, together with Tom Woodard, made access to San Basilio possible. Likewise, my return to San Basilio with colleagues David Backus and Jorge Ledesma-Vázquez in March 2017 and later in April 2019 with my son Erlend Johnson was hosted by Norm and Tom, as well as videographer Erik Stevens. Members of the Loreto Explorers Club accompanied me on trips around Isla Danzante, Puerto Escondido, and Tabor Canyon during my 2017 and 2019 visits that figured prominently in the design of chapters 6 and 7.

Correspondence with readers of my previous books on the geology of Baja California were vital to the outcome of this project. Foremost, I am indebted to the late Henry Ellwood of Grand Lake, Colorado, who regularly peppered me with questions regarding his observations in the Gulf of California during extended fishing trips spread out over many years of winter cruising on his boat *Cazador* (Spanish for "Hunter"). Hank and his spouse, Natalie, retired from their partnership in a Colorado law practice and made a second home for themselves plying the Sea of Cortez. In particular, Hank volunteered to explore the waters around Isla Santa Cruz and Isla San Diego on my behalf and provided extensive photographs. Those are spots I had long desired to visit, but never quite got around to doing so. Featured in chapter 8 is a small sample of Hank's photography, which Natalie freely made available for my use. We never had the opportunity to meet in person, and I am sadly deprived by the loss of such a valued correspondent. Correspondence with Ginni Callahan (proprietor of Sea Kayak Baja Mexico with headquarters in Loreto) provided additional crucial insights on the geology of Isla San Diego.

Artists Donna and Tom Dickson (former residents of Loreto now living in Puerto Vallarta) introduced me to the Pleistocene shell deposit in the hills behind Puerto Escondido, which they discovered on hikes and explored after reading my 2002 book (*Discovering the Geology of Baja California*). We carried on a lively correspondence about the place, and it gave me satisfaction to know that the book empowered those with an underlying curiosity about nature to apply geologic concepts on their own initiative. Donna and Tom's input is recognized in chapter 6 of this book. Likewise, the nature videography of Loreto resident Erik Stevens was influential in the book's treatment of storm deposits out at San Basilio and nearby Ensenada Almeja (chapter 4). Bill Burley, another longtime correspondent and Baja California enthusiast, took the time and effort to subject an early draft of the manuscript to a close reading, making numerous corrections to my errors in the citation of geographic place names.

Little did I know at the time that one of my students participating in an excursion to Baja California in the mid-1990s would be inspired not only to make secondary education his profession, but to exercise his devotion as a science teacher in the town of Todos Santos, Baja California Sur. Renewal of my acquaintance with Thomas Ekman, middle-school science teacher at the Sierra School in Todos Santos, has been most gratifying. College professors often are said to dwell in ivory towers, where students are eager to receive their teachers' wisdom. The challenge of teaching students at the seventh- and eighth-grade levels is demanding, if not entirely different by comparison. Tom showed me that it takes a gifted teacher to capture the imagination of younger students, whose attraction to science otherwise easily might be diverted. My visit to Todos Santos in 2017 was organized by Jack Clise, who enlisted my participation in the community's annual lecture series in the natural sciences. Jack went beyond the usual duties of a host by personally escorting me on a tour across the Sierra de la Laguna as described in chapter 10. The excursion gave me the pretense to cast a hard look across the wide Pacific Ocean from the end of the Baja California peninsula, taking into consideration the basin's growing storminess and its effect on Tom Eckman's students growing up in Todos Santos.

Last but not least, I am indebted to my adult son, Erlend M. Johnson, for his assistance with field work at Ensenada Almeja and Puerto Escondido, the results of which are shared in chapters 4 and 6. The work was arduous and I could not have done it alone. No less gratitude is due to my spouse, B. Gudveig Baarli, who endured without complaint my many seasons of leave-taking for adventures in Mexico's Baja California. Perhaps her forbearance has something to do with the marked improvement in my temperament at my home-coming that is slow to dissipate until the next possible leave-taking. She, too, also shared in more than a few ventures to this special land and played an especially important role in our visit to Cabo Pulmo (chapter 9).

Baja California's
Coastal Landscapes
Revealed

1

Global Warming and Forewarnings from the Geologic Past

BOOM!!! The entire glass wall of the lobby EXPLODED with glass, pieces of building, everything flying to the other side of the lobby like an explosion in an action movie.

—*Storm chaser Josh Morgerman, midnight, September 14, 2014*
FROM CABO SAN LUCAS DURING HURRICANE ODILE

NAVAL RESERVISTS SOMETIMES are called "summer sailors," because their active-duty training rarely takes them to sea during the stormy winter months of the year when otherwise employed ashore with civilian jobs. That moniker might be applied to me as a geologist, having never set foot in Mexico's Baja California peninsula during the annual hurricane season from September through early November. At some point in their professional lives, geologists who claim to understand the physical processes shaping Planet Earth aspire to the adrenalin-charged opportunity to witness an active volcano, to experience an earthquake, or to flee from a tsunami. Indeed, the field-hardened among us relish telling all who might listen about any such encounters based on personal experience. In terms of damage to infrastructure, Hurricane Odile entered the record books as one of the most violent storms to strike the Mexican state of Baja California Sur (Muría-Vila et al. 2018). Issues regarding the stability of roads, bridges, harbors, and other constructions, including commercial buildings and private dwellings, are of much concern to the civil authorities, not to mention the potential for loss of human life. Modifications to coastal features such as beaches and rocky shores as a consequence of storm impact are of equal interest to physical geographers and many geologists.

My career as a college professor kept me closely bound to a classroom for nine months of the year, with summer vacations free for field studies. The summers are too hot for fieldwork anywhere in Baja California, even along the cooler shores. In my case, I managed to make a month-long excursion to Baja California each January over an unbroken span of nearly three decades during what my home institution celebrates as a "Winter Study Period" between the Fall and Spring semesters. Often, I escaped to the peninsula for another two weeks during Spring Break. Cumulatively, I invested something on the order of three years living in Baja California while devoting myself to the study of its geography, geology, and paleoecology. Not once did I experience a rainstorm or gully washer of the kind called a *chubasco*[1] by the inhabitants.

Those with any experience navigating the highways of Baja California are familiar with the road sign that alerts the driver to a *vado*, or sizable dip in the road where it crosses an entrenched *arroyo*, or normally dry streambed. The margin of the streambed is fitted with a post marked off in meters to indicate the depth of water during a flood. Wide streambeds are traversed by concrete bridges, but the roadway sits on the streambed in smaller channels. These are easier to restore after a flood, when debris may be cleared by heavy equipment. Given fair warning, few are foolish enough to ford a stream under flood conditions, even when driving a heavy truck. The closest I came to experiencing a rainstorm was during a visit to Las Tres Virgenes, the volcano district located west of Santa Rosalia. During a January 1998 visit to Punta Chivato, my friend Marge Summers offered to take me on a tour to the site where drilling was being done to establish a sustainable system for generating electricity from a hydrothermal source. On reaching the end of the road, below the summit of the highest volcano with its peak at an elevation of 6,300 ft (1,920 m), big rain drops began to splatter against the vehicle's windshield. Without apology, Marge turned the car around and we raced back to the paved highway. Crossing numerous *vados*, the 25 miles (38 km) between Santa Rosalia and the Palo Verde turnoff to Punta Chivato was completed in record time. Arriving back home, the sun was shining brightly above a clear, blue sky. It was a false alert.

Year-round residents at Punta Chivato, Marge and her husband, Jere, had good reason to be wary of rainstorms. Their lovely home on the Gulf of California near Punta Cacarizo (also known as Hammer Head Point) narrowly survived destruction in 1997 during Hurricane Nora. Like most tropical disturbances in the eastern Pacific Ocean, Nora formed out at sea southwest of Acapulco on mainland Mexico. It became a Category 1 hurricane on September 18, as it tracked northwest in the direction of the Baja California peninsula. The storm intensified into a Category 4 hurricane reaching a maximum wind speed of 135 mph (210 km/hr) on September 21, before making landfall near Bahía Tortugas midway northward along the Pacific coast of the Baja California peninsula. The system stalled over land, but regained strength on reaching the warmer gulf waters on the opposite side of the peninsula. Intense rainfall deluged a large swath of the gulf coast, including Punta Chivato.

As recalled by Jere Summers, Nora brought steady rainfall to Punta Chivato over an 24-hour period, during which the ground surface was quickly soaked. In short order, sheeted runoff began to flow down slope entirely outside the capacity of local arroyos and smaller gullies. It was not the wind that was so unsettling after the system downgraded to a tropical storm. What startled the occupants of the split-level home with its lower-story patio open to the shore was that water began spurting out from electrical outlets in the downstairs wall built against the hillside. Essentially, the house behaved like a dam against the downward flow of ground-saturated water blocked by the building. Electrical outlets were the only available release for water ponded behind the wall. Reacting quickly, Jere ran to his garage workshop and grabbed a 10-pound sledge hammer. Racing back downstairs to the dayroom with its door open to the outside patio, he smashed a hole through the cinder blocks above one of the electrical outlets. The resulting gap released a torrent of water much like the opening of a fire hydrant. With the patio door now open, the gushing water flowed across the floor of the day room to exit outside across the patio. Without fast action to remedy the situation, the back wall of the structure was sure to have buckled under the pressure from the water behind it. Some part of the home's upper level was likely to have fallen into the void below. During the preceding dozen years that the Summers lived more-or-less full time at Punta Chivato, they had experienced nothing like it.

When I first heard their story, I regarded it as an odd mishap related to an erratic weather pattern, but I also understood why Marge insisted on racing home from Las Tres Virgenes to avoid being stranded in Santa Rosalia. During those early years of my January forays to Baja California, I was likely to be asked by my colleagues back home in Massachusetts if I had enjoyed my winter vacation in the sun. I returned home with a farmer's tan on my face and arms and the students who accompanied me for fieldwork had good things to report about our mapping projects focused on Pliocene and Pleistocene limestone formations. On its own account, limestone is normally diagnostic of a tropical to subtropical setting, and the limestone exposures around Punta Chivato close to 27° north latitude include former coral reefs that thrived in gulf waters appreciably warmer than today.[2] My visits to the Sea of Cortes were exceedingly pleasant, aside from those few days when a strong winter wind funneled down the gulf from the north and pushed white caps across the water surface. Now, as then, my overall impression of the place is that it mirrors a serene Mediterranean setting with all the advantages of stunning scenery and none of the drawbacks of over-crowding from too many people.

The next occasion during which I was forced to admit that the Sea of Cortes was not reliably placid year-round was during a visit to Loreto, some 80 miles (130 km) southeast of Punta Chivato a few months after Hurricane Marty struck on September 19, 2003, following landfall at the southern tip of the Baja California penin-sula the previous day. The system's counter-clockwise rotation as it migrated northward up the gulf spun out storm bands with the strongest winds and wind-driven waves in the so-called right-front quadrant. In a well-aimed blow, the storm landed a knockout punch against the northeast corner of the town's artificial harbor with its rectangular wall reinforced by huge boulders. The gap was surgically precise, blasting open a passage wide and deep enough to admit a cabin cruiser. On viewing the scene of destruction, one could only marvel at the force of nature required to accomplish such a blow. A year later, the harbor walls were rebuilt and reinforced with yet larger blocks quarried from the mountain range west of town.

During a visit to Loreto in January 2004, time was allotted for an excursion to Punta El Bajo at the end of a gravel road facing across a channel to Isla Coronados some 5.5 miles (9 km) north of town. Hiking the shore immediately south of the point popular with townspeople for picnicking and fishing, our group spotted something we'd never witnessed during previous visits. Thousands of rhodoliths the size of softballs were spread out like a white, lumpy coverlet across the narrow beach between low limestone cliffs and black cobbles exposed by the low tide. The color contrast was riveting to the eye: the black of igneous cobbles in the intertidal zone against the white-bleached remains of spherical-shaped coralline red algae left high and dry in the supratidal zone (figure 1.1). Composed mainly of calcium carbonate ($CaCO_3$), the unattached algae roll about on the seafloor under the influence of gentle waves and currents at a depth locally between 5 and 13 ft (1.5 to 4 m). All the while, the living alga continues to grow in diameter through concentric layering of calcium carbonate stimulated by photosynthesis under the strong sunlight penetrating shallow waters. The extensive rhodolith bed around an islet in the middle of the passage between the mainland and Isla Coronados was known to exist due to a mapping project carried out by divers from the Moss Landing Marine Laboratories in California.[3]

It became apparent at once during our January 2004 visit that the same storm responsible for knocking a hole in the Loreto harbor was the cause for so many rhodoliths having been swept ashore from the Coronados Channel. The same storm bands with winds roaring east to west from the hurricane's rotation as it moved northward into the Gulf of California had pushed larger-than-normal waves across the passage. For me, it was the start of a growing realization that hurricanes could have an outsized impact both on human communities as well as the continuity of nature around the Baja California peninsula. The death scene of bleached rhodoliths (red to maroon in

Figure 1.1. Supratidal zone (white) covered by bleached rhodoliths (spherical-shaped coralline red algae) near El Bajo, washed shoreward by storm waves from Hurricane Marty that struck the Loreto area in September 2003. The extinct Pleistocene volcano on Isla Coronados rises in the background. Photo by author.

color as living members of the plant division Rhodophyta) offered an opportunity to quantify the volume of degraded rhodoliths necessary postmortem to replenish the white beaches and sand dunes on the facing shore of Isla Coronados (Sewell et al. 2007).

During subsequent travels around the gulf, I became more aware of the scale of alterations to the physical landscape caused by major storms. Starting in January 2008, the first of several expeditions was conducted on Isla Cerralvo in the lower Gulf of California, east of La Paz. The initial visit was prompted by the prospect of locating a Pliocene limestone deposit formed almost entirely of rhodolith debris.[4] Following a vague description in the older scientific literature, those deposits were successfully rediscovered and studied (Emhoff et al. 2012). As it turned out, however, the same types of limestone formations so well exposed around Punta Chivato and Isla Coronados are underrepresented on Isla Cerralvo. What left a strong impression and helped to explain why the place was unsuited to the development of living rhodolith beds in clear seawater was the island's history of storm impacts. A review of storm tracks (figure 1.2), shows that Isla Cerralvo sustained direct hits from storms that passed directly overhead dating from Hurricane Fausto in 1996 to hurricanes Marty and Ignacio in 2003, to Hurricane John in 2006 and Lorena in 2019. Essentially, an island covering 52 square miles (135 km²) is being hollowed out from within by the action of rainwater dumped by episodic storms flushing arroyo sediments out to sea. Radiating outward to the island's perimeter, as many as 39 tidewater deltas could be mapped and the descending angle of their associated streambeds measured (Backus et al. 2012). The hurricanes known to have struck the island represent only the few tracked during the onset of satellite capabilities used by meteorologists in the modern era. Clearly, the granite-dominated island experienced a longer history of erosion related to the passage of hurricanes in the distant past, stretching back millions of years.

My visits to Isla Cerralvo evoked a fantasy to experience a hurricane from the safety of a bunker on a limestone ridge high above one of the tidewater deltas. Truth be told, other than witnessing the howling wind said to sound like a freight train, visibility of the flooding arroyo below surely would be blurred by torrential rainfall. Geologists don't really need to experience the force of a hurricane, because they can appraise the aftermath months or sometimes even years after a storm has impacted an area normally left arid for long intervals. An idea of the coarse mix of sediments available for transport by rushing water is found in a braided stream channel carved into one of the arroyo streambeds during late-stage flow some 165 ft (50 m) above sea level on the west side of Isla Cerralvo (figure 1.3).

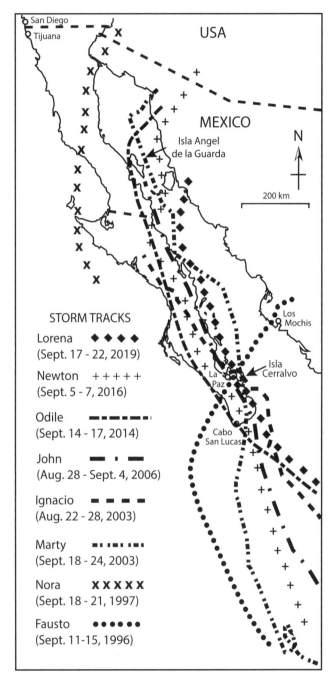

Figure 1.2. Historical storm tracks of the last seven hurricanes to impact the Baja California peninsula. Original drawing by author.

Figure 1.3. Valley floor of an arroyo streambed above the west coast of Isla Cerralvo in the lower Gulf of California. The island has sustained direct hits from several hurricanes. Photo by author.

Streambeds and their associated deltas up and down the gulf shore of Baja California increasingly took ahold of my consciousness as a topic of fresh interest. Not that I lost any passion for the diverse fossils preserved in limestone layers that accrued during the gradual separation of the Baja California peninsula from mainland Mexico beginning more than five million years ago, but the power of big storms representing fleeting instants in time by comparison had me question their effect on the resilience of marine life in shallow coastal waters. In the aftermath of Hurricane Odile, the town streets of Loreto remained awash in the seepage of ground water for months after the storm struck in September 2014. Loreto had suffered through many floods prior to Odile. Attending the mission church during one of my earliest visits to town, I spotted a faint line showing through the whitewashed walls of the naive, where the high-water mark of a long-ago flood left its trace. The civil authorities in Loreto always understood that their town was vulnerable to flooding, but after Odile they took action on a grand scale to

Figure 1.4. Levee system that entrains the normally dry streambed leading to the Loreto delta. The streambed is 675 ft (205 m) wide in this view. The northern part of Isla del Carmen across the Carmen Channel appears on the horizon. Photo by author.

safeguard infrastructure from future storms. The result was construction of a major levee system that entrained the Loreto delta and the lower river course feeding into it behind abutments much like the massive harbor walls (figure 1.4).

Before the project's completion, I scouted the stream's natural embankments in order to better understand how sediments transported by floodwater from the neighboring mountains make their final approach to the sea. A classic system of foreset gravel beds steeply inclined seaward was found in cross section where parts of the bank had been cut away by later torrents. Sediments from the greater Loreto system are smaller in diameter and better sorted than those clogging the many small arroyos on Isla Cerralvo (figure 1.3). Differences in streambed gradient are an important factor, as the Loreto system is remarkably level for the last 5 miles (8 km) on its path to the sea. The Loreto scenario exhibits a more orderly shift of its bedload to the delta in thin layers that tumble one over the other in serial cascades on a steady march to the ocean. There exists a living, organic world that has a fossil record showing to all who would pay attention the evolution of a truly great feature like the Gulf of California. Likewise, the gulf's geography includes physical appendages in the network of coastal arroyos that reveal a history of inorganic growth with a life all its own. Hence, the distinct organic and inorganic worlds of the gulf region meld together under the influence of storms that were certain to have disturbed the region from its earliest origins.

A limited number of historical storms on the Baja California peninsula and adjacent Gulf of California is recorded by a listing of some ninety-five events or tropical cyclones back through the 1950s.[5] Based on these data that include two storms as late as 2019 and 2020, some notion of changing climate patterns over the last nearly 70 years is hinted by a bar graph organized with frequencies registered by 5-year intervals from 1950 onward (figure 1.5). The graph draws a distinction between the period from 1950 to 1979, when there were fewer storms, amounting to less than one per year, and the period after 1980, when the number of storms almost doubled. Occurrences of Hurricanes John (2006), Odile (2014), Lorena (2019), and Genevieve (2020) are marked in this scheme as four of the more recent events. The existing historical record of big storms may be flawed by recognition of too

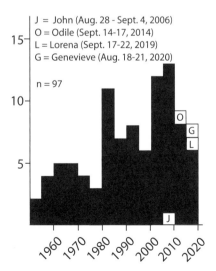

J = John (Aug. 28 - Sept. 4, 2006)
O = Odile (Sept. 14-17, 2014)
L = Lorena (Sept. 17-22, 2019)
G = Genevieve (Aug. 18-21, 2020)

n = 97

Figure 1.5. Bar diagram for the Baja California peninsula and Gulf of California showing the frequency of hurricanes or tropical depressions in 5-year intervals. Hurricanes John, Odile, Lorena, and Genevieve are shown in rank according to their order of arrival.

few events before the advent of satellite coverage. Even so, it remains a serious question whether or not hurricanes have increased in frequency and intensity over recent decades during which the rise of carbon dioxide in the atmosphere rose to 415 parts per million.

I seldom follow the weather news on a regular basis, but no matter where I might be, those who do are sure to alert me when a major storm threatens Baja California. The 2015 hurricane season was especially active over the eastern Pacific Ocean (Avila 2016), and the biggest storm of all was Hurricane Patricia, which rapidly formed as a Category 5 hurricane on October 23 of that year. Wind speeds reaching 215 mph (346 km/hr) and a minimum central pressure of 879 mb made it one of the largest ever recorded in the Pacific basin and the strongest ever to strike Mexico. For a time, it looked as though Hurricane Patricia was on track to strike Baja California. Dire warnings were issued for that part of the country. Due to an early and unexpected turn to the east, the center of the huge storm made landfall as a Category 4 hurricane on the Jalisco coast of Mexico far below the peninsula's southern tip. With a diameter of 1,500 miles (~2,400 km), the storm's outermost bands already swept across the opening to the Gulf of California. Had the storm continued northward and gained strength from the warmer gulf waters, damage to the infrastructure of Baja California and the potential for loss of life was likely to have surpassed the impact of Hurricane Odile only a year before. It was a close call for the region and a cautionary warning for ever-larger and stronger storms yet to come.

The NASA mission for Global Precipitation Measurement (GPM) links an international network of satellites to provide detailed observations in real time on a global scale regarding Planet Earth's rainfall and snowfall.[6] Information is collected for the viewing of precipitation patterns every 30 minutes, and the satellite feed may be seen as if through a single omnipresent eye, akin to watching a movie. The contrast of action and inaction can be mesmerizing to witness. Some parts of the world are rarely if ever touched by rainfall, such as the Atacama Desert in northern Chile, the central Sahara of North Africa, or the Gibson Desert of Western Australia. The skies above these places on the GPM montage remain clear. Elsewhere, the skies above the Amazon River basin of South America or the Congo River basin of Africa are perpetually cloudy with the buildup of daily rain-storms. For riveting action, however, nothing beats the commotion of swirling hurricanes or typhoons crossing the Atlantic, Pacific, and Indian Ocean basins during the latter part of each year.

Atmospheric and marine scientists broadly agree that human-induced climate change is related to the growing severity and frequency of hurricanes and typhoons around the world. The big storms are most prone to emerge over tropical oceans, where the sea-surface temperatures (SSTs) exceed 79°F (26°C). The higher the SST in any given area and the longer it is sustained, the more fully requisite conditions are in place to spin off a succession of storms. According to a global study by Oliver et al. (2018), evidence indicates that longer-lasting and more frequent marine heatwaves have grown globally by 54 percent in terms of annual marine heatwave-days since 1925. An independent survey by Kossin et al. (2020) suggests that hurricanes have increased in strength by 8 percent each decade over the last four decades. The scenario includes the general pattern of El Niño years in the equatorial Pacific Ocean, during which elevated SSTs episodically give rise to storms and above-average rainfall affecting the Pacific coast of North America. With ongoing global warming, more severe storms are expected as El Niño years lengthen in duration and spawn powerful storms like Hurricane Patricia.

Oceanographic evidence points to permanent El Niño conditions having been in place during the early part of the Pliocene Epoch from 4.5 to 3.0 million years ago (Wara et al. 2005). Aside from the relatively brief interglacial episodes of the last several hundred thousand years, the Pliocene Warm Period is the most recent interval of geologic time with a climate interpreted as warmer than today. Based on analysis of ocean cores containing microfossils of amoeba-like organisms called foraminifera that possess a calcareous outer shell (or test), it is possible to gauge differences in water temperatures between the former surface to a depth of more than 300 ft (~100 m). The SST along the equatorial Pacific Ocean was elevated with little east-west variation, much like a contemporary El Niño event but far more persistent over time (Wara et al. 2005). Moreover, the layer of warmer surface water persisted to a greater depth compared with today, meaning that the equator's thermocline was better fortified against perturbations. Assuming the interpretation of microfossil data is correct, the Pliocene Warm Period portends a kind of vision for what lies ahead on Planet Earth as CO_2 levels increase and present trends in global warming continue unabated.

As with any high-stakes forecast looking to the future, its accuracy becomes ever more focused with the conformation of supporting evidence from independent sources. For example, it is pertinent to ask to what extent the oceanographic evidence for the Pliocene Warm Period agrees with land-based geologic evidence from places where increased storms and related rainfall might be expected. The rock record from the Baja California peninsula and its associated gulf islands is surprisingly complete not only for much of the Pliocene, but also for parts of the following Pleistocene and the last 10,000 years of the Holocene. With a model in hand for what happens to an island like Cerralvo repeatedly hit by modern-day hurricanes, it remained only to search for similar traces of much older flooding and significant changes to coastal features as a test for the concept of the Pliocene Warm Period and its direct bearing on the Baja California peninsula. Moreover, it is relevant to ask to what extent storms over Baja California may have abated during the later Pliocene and succeeding Pleistocene prior to today. In total, the span of time under consideration covers roughly 5 million years.

In desert places that lack rainfall but commonly sustain strong winds, the sediments and rocks are dominated by sand and sandstone that represent shifting sand dunes. Baja California has its share of coastal dune fields commensurate with its level of aridity.[7] For places with a regimen of daily rainfall typical of equatorial rainforests in South American and Africa, river mud and silt are the prevailing sediments. These will eventually compress to make shale and siltstone. In marked contrast, a chaotic mixture of boulders, cobbles, and gravel issues from zones where heavy rainfall episodically impacts higher elevations akin to the central spine of the Baja California peninsula. The individual rock clasts contributing to such a mix all derive from the parent rocks exposed in the landscape. For Baja California, the predominant source rock for the generation of a poorly sorted conglomerate will be igneous rocks dominated by andesite, basalt, and granite.

Itinerary for Selected Rambles on a Peninsula Tour

The chapters in this book trace a personal transformation from a geologist with an affinity for limestone associated with fair weather conditions to one who sought out the inorganic rocks and deposits connected with foul-weather floods and stormy seas. That is not to say that the mood, herein, is correspondingly dark. All explorations were undertaken during January or March under sunny skies with scarcely any cloud cover. The Baja California peninsula remains for me a singular place for adventure and intellectual enlightenment, where all rocks (including limestone) are accessible in an open desert landscape bordered by seas on opposite coasts. The tour amounts to a hopeful journey in search of linked stories and events across geologic time (figure 1.6). These stories may well pertain to our future as informed stewards of the whole Earth.

Era	Period	Epoch	Absolute age
Cenozoic	Quaternary	Holocene	10,000 yrs
		Pleistocene	2.59 my
	Neogene	Pliocene	5.33 my
		Miocene	23.03 my
	Paleogene	Oligocene	33.9 my
		Eocene	55.8 my
		Paleocene	65.5 my
Mesozoic	Cretaceous		145.5 my
	Jurassic		199.6 my
	Triassic		251.0 my
Paleozoic			542 my

Figure 1.6. Geologic timescale with an emphasis on epochs from the Cenozoic Era.

Staged from northwest to southeast parallel to the peninsular axis (figure 1.7), the eight chapters that follow feature coastal spots with the opportunity for a guided ramble through scenic terrain. Each has a not-so-hidden story to tell regarding the land's challenges shaped by climatic factors playing out over a few million years. Chapter 2 (locality 1) concerns the basaltic monolith of Pliocene Volcán Prieto south of Puertecitos and the Holocene delta and salt lagoon on its flanks. Chapter 3 shifts south to Punta Ballena and the massive Pliocene delta that issued from a gap in coastal granitic uplands. Chapter 4 skirts around Punta Chivato and Bahía Concepción (covered in previous books) to focus on the extraordinary San Basilio area with its Pliocene volcanic islets and Holocene storm deposit. Chapter 5 brings us to the east coast of Isla del Carmen, the third largest island in the Gulf of California, where a cross-section through an enormous Pliocene delta system is exposed in sea cliffs. Chapter 6 expounds on the secrets of Puerto Escondido and neighboring Tabor Canyon (locality 5), famously visited by Ed Ricketts and John Steinbeck in 1940 during their voyage on the *Western Flyer*. Chapter 7 examines the geomorphology of Isla Danzante. Chapter 8 indulges in a dream-trip to the granitic islands of Santa Cruz and San Diego. Chapter 9 contemplates the rocky shores and fossil deposits bordering the gulf's only coral reef. To conclude, the epilog in chapter 10 brings us across the mountain divide at the southern tip of the Baja California peninsula, ending at Todos Santos with a view across the Pacific Ocean all the way to the other side where the island of Taiwan sits off the China mainland.

During our rambles, we traverse way stations on a multi-million-year passage from the deep past to the near future of lands bordering the great Pacific basin. What we may learn about changes affecting the Baja California peninsula echo with meaning across the many islands scattered throughout the great ocean and far-off continental shores on the opposite side.

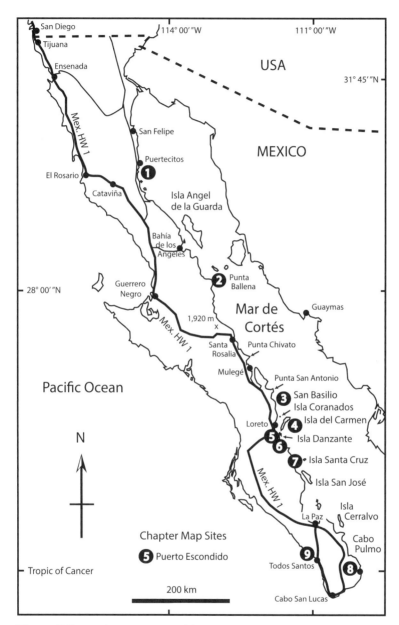

Figure 1.7. Localities in Baja California and associated islands treated in this volume. Map by author modified from Johnson (2014). Note: locality numbers do not correspond to chapter numbers.

2

Volcán Prieto and Salt Pan near Puertecitos

Sea storms are rare along this desert coast.
But when they strike, they do with a fury
of tropical cloudbursts, savagely altering
the contours of the land.

—*William Weber Johnson, Baja California (1972)*

THE BLACKTOP ROAD leading south from Puertecitos unfolds like a smooth ribbon undulating across the land's barren contours, but snug against a pale-blue sea. It is still early in the cool of the day, and the profile of Volcán Prieto rises darkly on the horizon. Consulted long in advance, the big geologic map at a scale of 1:250,000 for the entire northern state of Baja California shows a brown lump nestled against a field of orange.[1] The brown is for the basalt of Volcán Prieto, assigned to a Quaternary age (i.e., less than 2 million years old). Other brown spots pepper the center of the map, far from the coast. According to a study by Stock (2000), the orange represents the Puertecitos Volcanic Province attributed to rocks cooled from swift-moving flows called ignimbrites that slide down the slightest of slopes on cushions of trapped gas emitted like puffs of air strangled beneath lava blankets. Italy's Mt. Vesuvius released such an ignimbrite, or glowing avalanche, in the year 79 CE that engulfed the Roman towns of Pompeii and Herculaneum before their citizens could hope to escape.[2] As for the Puertecitos ignimbrite, it entails a volcanic field more than one hundred times the footprint of Mt. Vesuvius, and it fed repeated eruptions over a vastly longer interval of time between 6 and 3 million years ago. No one witnessed those fearsome events, because the earliest humans to enter North America arrived across the Bering Strait from present-day Russia long afterward, during the last several thousands of years. Any living thing near enough to behold the intensity of the Puertecitos ignimbrite from a safe distance would want to decamp as quickly as possible.

What brought me here on a spring day in the company of my field partner, Jorge Ledesma, were enticements other than the massive strata of the Puertecitos ignimbrite. I knew the place only on the basis of a satellite image and a closer view from 30,000 ft (9,144 m) during a commercial flight from Loreto in Baja California (Lower California of Mexico) to Los Angeles in Alta California (state of California in the United States). I had helped to choose the twenty-six satellite images in the *Atlas of Coastal Ecosystems in the Western Gulf of California* (2009) to provide 100 percent overlapping coverage along the entire gulf coast of the Baja California peninsula. The goal was to provide a means of exploration for those attracted to the varied landscapes around the gulf and its many islands, not so easily reached by ground transportation.

As the final exercise in a semester-long course on the geology of Baja California, I asked each of my students to find and report on a particular spot that piqued their curiosity based on one of those satellite images. The winning presentation was for Volcán Prieto and its signature white spot that shined like a bright beacon from the old crater. Much of the research conducted jointly by the American and Mexican students under our supervision during the previous years was focused on limestone. We had been rather successful in finding limestone on some of the more remote islands in the Gulf of California. The satellite images, which differentiate limestone from the surrounding rocks as white or typically beige patches, led us directly to those places of greatest interest. Thereby, the images saved valuable time both in advance planning and ground efforts. The beige dot sitting on top of Volcán Prieto was small, but intense. Moreover, a decent road gave easy access to the site. Being able to pick out the place during a commercial flight (plate 1) only heightened my determination to get there.

If there had been any prospect that marine water breached a dormant volcano and flooded its crater, it would be intriguing. For example, the caldera of a volcano on the island of Sal in the Cape Verde Islands of the North Atlantic is the site of a commercial operation for the extraction of salt from the ongoing infiltration of marine water enclosed by the crater walls.[3] But I also understood from the satellite image that the crater floor at Volcán Prieto sat more than 800 ft (~245 m) above sea level. Did the volcano have a history of tectonic uplift? A genuine salt flat resides on the south side of Volcán Prieto, and Jorge had been there during an independent trip. He assured me that living microbes similar to those we had discovered and compared to fossil stromatolites in the closed lagoons of Isla Angel de la Guarda (Johnson et al. 2012) were likely to be present at the side of Volcán Prieto.[4] Should we be able to obtain samples of organic material from the salt lagoon, a laboratory group at the University of Vienna stood ready to perform an analysis and identify the

Figure 2.1. View of Volcán Prieto, a young edifice, at the side of Mexico Highway 5 south of Puertecitos by 7.5 miles (12 km). Photo by author.

bacterial species present. Hence, we had two reasons to visit the area. Laden with the supplies necessary to collect and preserve the organic material, we came prepared with detailed instructions on how to use chemical reagents to complete our mission. Some 50 hours after landing in San Diego at the conclusion of a cross-continent flight, I stood with Jorge at the side of Mexico Highway 5, peering at Volcán Prieto's silhouette (figure 2.1). We would climb to the peak and solve the mystery of the white spot during the morning hours and only afterward proceed to the nearby salt flats.

FEATURE EVENT
Climbing Volcán Prieto

The topographic map for Volcán Prieto (map 1) shows a prominence rising more than 900 ft (275 m) above the adjacent Gulf of California. Our geologic map indicates a major fault that runs through the center of the structure from north to south. Mexico Highway 5 reaches a maximum elevation of 295 ft (90 m) along the western flank of the old volcano, saving much of the climb otherwise necessary from a lower elevation. The side access used during the paving of the highway provides a safe, off-road space to park our vehicle during the ascent of the western slope. The starting point promises a shorter and less precipitous climb than anywhere from within the dissected fault valley. The march begins, initially straight upward over ground strewn with rough-edged boulders, but thereafter making traverses from side to side to conserve energy with each step. Although sparse, the dominant ground cover is the brittlebush (*Encelia farinosa*), a member of the sunflower family that sports bright yellow flowers when in bloom. It is springtime, but the ground is parched, and the plants show only their silver-gray leaves attached to woody branches that rise above knee height. Each plant is radial in design, but well-spaced away from its nearest neighbor. The open ground allows for easy passage among the bushes.

We stop midway up the slope to retrieve a rock hammer from a daypack and break open a block of stone lying flat on the surface. On the outside, it has a brownish-red patina that resulted from long

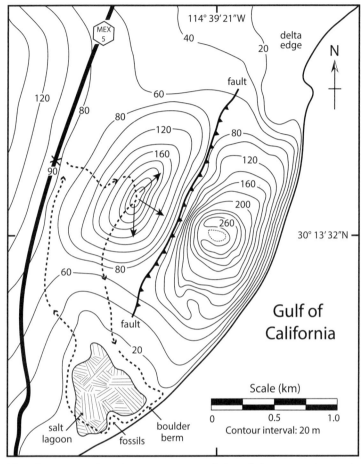

Map 1. Topography of Volcán Prieto and the salt lagoon on its south side. Dashed line with look-out points (arrows) shows the featured hike for this locality.

exposure to the elements. In contrast, the freshly broken surface exhibits a dark gray color. A quick look with a hand lens at 10x magnification confirms that no sparkling bits of silica are present in an otherwise fine-grained rock. As claimed by the geologic map, the rocks of Volcán Prieto are formed by basalt. The halt is a good excuse to catch one's breath from the climb, but there remains an underlying urgency to continue upward for the expected reward at the top of the rise. It had been a few years since my attention was drawn to the telltale satellite image. The long hours on the cross-county flight a couple days before may as well have been years. A nagging suspense pushed me to resume the climb with increased vigor. It had been a long wait for any kind of definitive answer.

The first surprise revealed from the crown of the hill is unexpected, although my earlier aerial reconnaissance (plate 1) should have prepared me better for the sensation of a view from a small aircraft at a low altitude. In full view to the northeast sits a large delta system with a curved outer fan stretching for at least 3 miles (~4.75 km) where it meets the gulf waters. Crossing the delta near its apex at the mouth of Heme Canyon, the road elevation is 165 ft (~50 m) above sea level. From such a low position looking seaward across the fan, it is impossible to recognize the delta's three-dimensional form. In fact, the gradient from the road outward through the central distributary channel to the sea amounts to a mere 2° gradient. Here from the heights of Volcán Prieto, the characteristic delta shape (which takes its name from the Greek letter Δ) exhibits an obvious bulge pushing into the Gulf of California.

It is the first of the day's enigmas with regard to climate. The size of the delta is no match for the mighty Colorado River delta to the north, but the Playa Costilla delta is a sturdy pigmy. In total area, it covers no less than 3 square miles (8 km²), but immerges from a narrow canyon that extends inland for only 9 miles (14 km). The delta is largely free of vegetation, and it reveals multiple distributary channels well entrenched in its surface. The Colorado River is one thing, tracing upstream from its mouth for 1,450 miles (2,333 km) into the heart of the Great American Southwest. When the Colorado River ran free unobstructed by river dams, it carried an immense sediment load, much of which found its way downstream to the delta. The normally dry streambed within Heme Canyon is extremely short, but provides a disproportionate amount of river gravel transferred to the gulf's edge. The existing delta could only be the result of running water, lots of water, that regularly flushed a sizable volume of eroded sediments out of the canyon and through the multiple distributary channels leading to the delta front. It might be assumed that the present-day delta is an artifact of the distant past, when rainfall may have been more plentiful than the 2 in (5 cm) of annual average rainfall typically registered for this driest sector of the Baja California peninsula.[5]

As prominent as it is, the local delta seems at first like a vexing distraction. Crossing a short distance south along the summit, the long-sought view into the volcano's crater is at last realized. An opening to the interior sits slightly below eye level less than a half mile (750 m) away (map 1). I remove my day pack, take a seat on one of the flat stones, and find the binoculars for a closer look. As determined by Kozlowski et al. (2018), the crater rim has a circumference of about 1,150 ft (350 m), but is open on the west side due to a breach in the walls. The flat crater floor sits at least 50 ft (~15 m) below the crater rim, and the beige sediments are readily distinguished from the dark, bowl-like margins that enclose it on three sides. Sitting here, it requires some pause for reflection to process all that is in plain view. Some of the beige coloration radiates outward from the core in streaks that rise ever so slightly to higher ground, but still far below the crater rim. The thought begins to take hold that the beige deposit has the look of a dry lake bed, with the adjoining streaks representing rivulets converging on the center. Moreover, there remains just enough of a low western rim to prevent the beige material from overflowing outside the crater. No trace of coarse layering is evident, and the pattern is inconsistent with limestone. Nor does it appear like a salt deposit, which might be expected to assume a purer white overtone.

Running water is the ingredient needed to make sense of the puzzle, but not as much water as required to construct the nearby delta system. Runoff from rainfall over a much larger area comprising the watershed around Heme Canyon accounts for the long-term development of the Playa Costilla delta. The watershed for the Volcán Prieto crater is miniscule by comparison, amounting to barely 700,000 square feet (65,000 m²). Unlike the coarse gravel deposited in the delta system, the beige material on the crater floor comprises fine-grained sediments consisting of silt and clay. To accumulate such a deposit, a cloud-burst with steady rain would need to make a direct hit on the crater, perhaps stalled over the area for a protracted amount of time. The rain would scrub the crater walls, removing dust and silt to be carried off in a kind of liquid slurry to the center.

Although not a thick lake deposit, accumulation was sure to have resulted from the repetition of rainstorms over time. How could this make sense, given the average rate of annual rainfall in this parched, desert environment? Not only is a substantial amount of water required, but downpours had struck a small target with the unfailing precision of an expert marksman.

Tempting as it is to cross over the fault valley and enter the crater, our agenda required the remaining afternoon to explore the salt lagoon a mile (1.6 km) away (map 1). Viewing points along the descent beginning from the south end of the summit provide an eagle-eye perspective on the triangular salt pan (figure 2.2) that reflects the whiteness of a frozen, snow-covered lake not unlike those in the dead of winter in the Berkshire Hills of my Massachusetts home. The lagoon is well isolated from the adjacent Gulf of California by a long berm composed of basalt boulders. The berm acts like

Figure. 2.2. Salt lagoon on the south side of Volcán Prieto, sheltered behind a strong berm with boulders eroded from the volcano's exposed sea cliffs. Photo by author.

a fortified flood wall, more than a mile (1.6 km) in length, keeping the marine waters of the gulf from pouring directly into the lagoon. Around the sides of the lagoon, a pale-blue refection suggests the presence of shallow, open water. Halfway down the slopes of Volcán Prieto, the topography evens out, and the descent to the near side of the lagoon is easier to negotiate. Close to sea level along the margin of the lagoon, the next third of a mile (0.5 km) brings us to the boulder berm attached to the south corner of Volcán Prieto. The berm tops out less than 12 ft (~3.7 m) above mean sea level at the far north end. We have rambled through a little more than half the day's hike, and the refreshing onshore breeze at this spot offers as good a place as any to sit and enjoy lunch.

Even during a midday repast, it is hard for the truly dedicated to fully disconnect from the surrounding environment on full display. Earlier in the morning, we verified that Volcán Prieto is formed by basalt, although the outer varnish of exposed stones is brownish red in coloration. Here along the berm looking southward, the dominant boulders are clean basalt that register as flat-gray in color. Therefore, it is all too shocking to see a splash of orange cobbles that trace out a pathway down the center axis of the berm some 30 ft (9 m) removed from the water's edge (plate 2). However enjoyable, no field lunch could stem the curiosity that overwhelms a geologist wanting to answer the question of the moment. What could these orange lumps be? When picked up, the answer is immediately at hand, because the cobbles are extraordinarily light for their size. They are composed of pumice, frothy and delicate in consistency with a density significantly less than that of water. Basically, pumice is a volcanic rock made of compound elements in silica (SiO_2) but full of air vesicles. Pumice floats in water. Like children smitten with a new toy, the adult geologists occupy themselves over the rest of the lunch hour by tossing the largest cobbles into the water and watching them bob up and down in the waves. The orange lumps look ever so much like the small, yellow rubber duckies one remembers playing with among the soapsuds during bath time.

It might be claimed that geologists are vulnerable to sudden behavioral shifts related to the Peter Pan principle. One might look like an adult, but many a grown-up geologist conceals a playful child inside. Playtime can last only so long, before more mature questions arise. Clearly, the pumice "duckies" floated into place during a high tide, probably pushed a bit further ashore by windy conditions. But where did the pumice come from? The answer is not directly obvious, as there were no outcrops of pumice to be seen anywhere during the morning's ramble. To get a better handle on the likely answer, it is necessary to fall back on the larger regional picture as seen from high above (plate 1). Stratified layers of the ignimbrite exposed within Heme Canyon are likely to include horizons laden with pumice. Indeed, one or more of the distinctly colored layers exposed in the canyon walls not far upstream from the road crossing could be the source of the pumice. Floodwater coursing down the canyon could pick up chunks of pumice, which thereafter were certain to be flushed through the distributary channels of the Playa Costilla delta. On entering the Gulf of California, the pumice duckies would be carried along the seafront off Volcán Prieto by longshore currents before arriving at and becoming stranded on the boulder berm. Thus envisioned, the scenario floats a pleasing argument in support of the notion that the delta is not inactive, but very much alive to present-day processes. When was the last time Heme Canyon flooded, and when was the last occasion when Volcán Prieto's crater received a fresh addition to its clay floor? The answer, it seems, might be available from meteorologic weather records in due course.

We must, however, resume our ramble and follow the orange pathway along the boulder berm to a point where we may descend to the level of the salt lagoon marked by an indentation oddly free from the white crust. Another surprise awaits, because we soon stand on a limestone shelf that projects out from beneath the cover of the boulder berm. The pale limestone is full of fossil shells belonging to a species of mollusk called a venus shell (*Chione californiensis*). The species is found alive today, widespread throughout the Gulf of California, and characteristic of an intertidal habitat often on mudflats but also living a little farther offshore.[6] The fossil is one of the more common representatives typical of Pleistocene deposits along gulf shores of the Baja California from the last interglacial epoch dating to about 125,000 years ago.

A similar Pleistocene limestone with abundant venus shells occurs in a comparable setting behind a coastal berm at Punta Ballena (see chapter 3), and the same fossils are commonly found in limestone deposits around Punta Chivato and Isla Coronados.[7] This particular ledge of limestone exposed inside the perimeter of the modern boulder berm is the most northern locality within the Gulf of California known to feature Late Pleistocene fossils from about 120,000 years ago. The interpretation is not so difficult. We need only place ourselves a little further back in time before the erosion of the basalt boulders that make the present-day boulder berm. Volcán Prieto surely existed at that time, but the closed embayment that now shelters the salt lagoon must have been open to normal marine water when the Venus shells were alive at this spot. Taking another look at our topographic map (map 1), it now seems perfectly evident that the volcano's crater is not at the center of the structure as it should be for what surely was a symmetrical structure. The central fault informs us that the volcano had a larger perimeter than now preserved. Moreover, the east face of Volcán Prieto marks a cliff line with a gradient steeper than anywhere else around the structure. The original slope on the seaward side of the volcano must have projected at a lower gradient farther into the Gulf of California before coastal erosion took its toll. Prior to growth of the boulder berm, the shallow embayment on the south side of the volcano would have been larger and entirely open to the sea. The fossil marine shells clearly attest to this former reality.

The present-day salt lagoon remains to be addressed, and our ramble takes us closer to the south corner of the deposit (map 1). Here, we gingerly walk out across the salt-crusted surface, in some places sinking slightly into the soft surface. Using the shaft of my wooden walking pole, I find that it can be pushed at least 2 ft (65 cm) into the sediments below. When the shaft is withdrawn, the small hole barely an inch (2.5 cm) in diameter fills with water. Dipping a finger into the water and touching it to my tongue, there is no doubt about the hypersalinity of the water. I need to flush my mouth with fresh water from my water bottle. Closer to the west margin of the lagoon, shallow pools of water remain open to the surface. Retrieving a knife from my day pack, I cut a shallow trench into the sediment not far off from the open water. The operation is done carefully to preserve a clean profile in the cut. Immediately below the salt-crusted surface there appears a distinctly red line of organic matter. Below that, within a space of 5.5 in (15 cm), there follow thin strands of green organic matter separated by intervals of wet clay. The colors are startling and unexpected compared to the black organic matter previously sampled by our team from a closed lagoon on Isla Angel de la Guarda (Johnson et al. 2012). But here we have struck "pay dirt."

Other tools and chemical solutions necessary to sample and preserve the organic matter for laboratory analysis are in the vehicle. With high spirits on a fine day with a southerly breeze in the air, we hike the 1.25 miles (2 km) along the highway's side access to retrieve our vehicle and bring it to the side of the lagoon. The work occupies the remainder of the afternoon, after which the return trip to San Felipe, 56 miles (96 km), completes the day (see overview map in figure 1.7). Multiple tasks are accomplished, and a celebratory dinner in town is in order.

LAB ANALYSES

Microbes include bacteria and other simple cells called archeobacteria that interact in complex communities to form structures with a long pedigree stretching back more than 2 billion years in Planet Earth's history. Geologists and paleontologists know these structures as stromatolites that typically occur as thin sheets, often in a dome-shaped structure or in flat layers. The layers are not unlike a lasagna, in which the marinara sauce can be compared to the organic matter and the alternating pasta layers represent inorganic silt or clay. Often, but not always, the living microbe cells are strung together in long filaments covered in mucus that protect the cells within from damaging ultraviolet light. The sticky filaments trap and bind fine sediments that accrue during tidal cycles or other changes in the environment, after which more bacterial filaments recolonize the surface. It might be said that stromatolites are akin to a bottomless lasagna dish that keeps adding new pasta layers and marinara sauce from the top, upward.

Only rarely is the organic marinara sauce preserved intact in fossil stromatolites. That is why studies of living microbes analogous to the fossil stromatolites are so valuable as a window on the past. Virtually nothing is known about the biodiversity that has vanished from fossil stromatolites, recognized only by their layered, inorganic structure. Yet those structures have a long geologic history that persisted into the later Pleistocene in the Gulf of California (Backus and Johnson 2014). The lasagna dish signified by the salt lagoon at the side of Volcán Prieta covers a surface area of 317 square yards (~265 m²) and extends to a depth of at least 2 ft (65 cm). How the biology of living microbial communities varies with depth is an intriguing question answerable only through sophisticated laboratory treatments of different kinds, including analysis of ribosomal RNA genes and lipids (fatty acids).

Biological samples collected from the salt lagoon in the shadow of Volcán Pieta were hand-carried across the world's busiest international crossing between Tijuana and San Diego at the conclusion of our spring 2016 expedition. There was a level of nervousness on my part about this crossing that might be regarded as an act of smuggling. Generally speaking, border officials are alert to the more egregious examples of cross-border exports, such as a vehicle with the skull and horns of a mountain sheep mounted on the front grill. My personal luggage contained suspicious-looking vials filled with cloudy liquids under seal and other packets filled with white, powdery material. On the Mexican side of the border, I stood patiently in line for several hours on a Sunday afternoon before I reached the lone border official on the United States side of the border. After asking me where I was from and why I was visiting Mexico, she told me she was a recent graduate of Mt. Holyoke College, not far from my home in western Massachusetts. "Welcome home," she said with a pleasant smile. I had yet to pass my luggage through the x-ray scanner.

Early the next day, the samples were repackaged and flown from San Diego to Vienna via an overnight delivery service. It required nearly two years for the scientific results to be published (Kozlowski et al. 2018), but it was worth the wait for the first detailed analyses of microbial communities from a habitat subject to extreme variations in water salinity that just happened to come from Mexico. Extracted from a pair of core samples 7 in (~17 cm) in length, evidence indicated the presence of twenty-five named bacterial phyla and eight archaeal phyla, in addition to several candidates for yet-unnamed groups. Lipid biomarkers from the same set of samples demonstrated that the cyanobacteria most often linked to fossil stromatolites were restricted to the upper levels within the cores, whereas anoxic forms lived deeper below the surface. In short, the length of core was alive with microbes from top to bottom, although different kinds were adapted to different conditions with depth. The number of closed lagoons with elevated salinities are plentiful along the Gulf of California. The salt lagoon at the base of Volcán Prieto is perhaps the most accessible, but a multitude of others await serious study in order to learn how bacterial life specialized for extreme habitats are able to spread from place to place.

THE CLIMATE CONNECTION

Few of the ninety-four hurricanes and subtropical storms known from the historical record to impact the Baja California peninsula since 1950 (see figure 1.5) made their way at full strength to the region around Puertecitos along the upper Gulf of California. In September 1976, one of the few big storms to transit the upper gulf region was Hurricane Kathleen, which struck southern California on the U.S. side of the border with wind speeds clocked at 99 mph (160 km/hr) as a Category 2 hurricane.[8] The last major rainfall to impact Heme Canyon, the Playa Costilla delta, Volcan Prieto, and the salt lagoon was from a downgraded Hurricane Odile in September 2014. Volcanic eruptions have long-since ceased, but otherwise none of the physical features in the coastal Puertecitos area are defunct relics of a weather system from long ago. The average annual amount of rainfall is very low, but large amounts of water fall as rain on the region, perhaps on a decadal basis. The canyon leading to the Playa Costilla delta floods, the delta discharges copious amounts of gravel (including pumice cobbles), the extinct crater at the top of Volcán Prieto turns into a shallow lake, and the salinity of the closed lagoon on the south side of the volcano is reduced by an influx of fresh water. Measured in human years, long intervals of drying time elapse between the big wettings. It is something of a paradox that the small salt lagoon sits so near the large delta, one a reflection of tremendous aridity and the other a manifestation of repeated flooding on a grand scale. Between them towers the edifice of the extinct Volcán Prieto with its intact central crater. In my dreams, I relish the opportunity to sit on the rim of the Prieto crater and witness a deluge of sustained rainfall!

3

Punta Ballena and the Pliocene Ballena Fan Delta

HURRICANE! And we are caught in it
—right in the middle of the Gulf—November 24 [1957]
—drifting helplessly toward the channel
where a 100-mph wind and a 20-ft
tide combined to convulse the sea.

—*Gene Kira quoting Ray Cannon, The Unforgettable Sea of Cortez (1999)*

NO SINGLE INDIVIDUAL was more responsible for promoting the romance and adventure of the Gulf of California than Ray Cannon (1892–1977), especially in regard to sport fishing. Inauguration of Mexico Highway 1 in 1973 made overland access to much of the peninsula and many points along the Gulf of California accessible to all who wished to drive its 1,563 miles (1,711 km) between Tijuana and Cabo San Lucas. Before then, the roads were unpaved and arduous, with better access to the gulf afforded from the Sonoran side. Cannon wrote popular columns for the *Western Outdoor News* and through that medium described his many excursions to an angler's paradise in places equally beset by solitude and great beauty. One of his proudest triumphs was to organize flotillas composed of private boats owned by Americans willing and eager to experience some of the prime fishing spots described in his columns. Normally, any accounts by Cannon of chaos on the water had more to do with fish boils that riled the waters and set off an explosive feeding frenzy throughout the food chain. Sport fishermen guided by Cannon were ecstatic to find themselves in the midst of such bedlam.

The 1957 fishing expedition consisted of four vessels: "17 to 21 footers equipped with twin out-board motors" that assembled on Kino Bay south of Isla Tiburon on the Sonoran coast. The intention was to motor across the gulf's midriff region roughly 124 miles (200 km) round trip to the opposite shores of Baja California near Punta Ballena and back again. The storm was unexpected, and the armada's participants attempted to shelter overnight halfway across in the lee of Isla San Esteban. Eyewitness narratives of big storms on the Gulf of California are few. This one could have ended badly. Cannon had finished his watch aboard his host's boat at midnight, but was awakened at 2:00 a.m. by noises unlike any he'd experienced before. A change in circulation brought wind gusts through a ravine spotted with caves. "The hundred-mph flow turned all the caves into screaming whistles; the sounds were louder than air-raid sirens, combined to produce a piercing, eerie pipe-organ effect" (Kira 1999, 158, quoting Cannon). One by one, the boats lost mooring anchors due to a rising tide, but survived the night after being lashed together with a larger vessel that miraculously happened to be in the same vicinity. Cannon relates that the region had not felt a storm of such intensity for the previous 20 years.

The aftermath of a mighty storm at sea may be measured in two ways as it pertains to any stretch of coastland within reach. The event weathered by Cannon and friends at Isla Esteban would have altered its rocky coast by direct wave action. Large rocks loosened and shifted by strong waves may pile up as coastal bolder deposits (CBDs) arrayed along the shoreline. At the same time, higher elevations above the reach of the waves may sustain erosion in ravines by way of rainwater runoff after quick saturation. Debris washed downslope toward the sea may coalesce separately as coastal outwash deposits (CODs). However minutely, the coastal caves on Isla San Esteban also are subject to enlargement by wind and lashing rain. The rest of Cannon's trip unfolded without incident, when the flotilla safely crossed the rest of the midriff on the following day and fished the gulf's western shores.

Punta Ballena and Bahía San Rafael on the gulf's peninsular coast (figure 1.7) are fascinating places on account of their extraordinary geomorphology. The true nature revealed by the shape of the land is not easily appreciated from a boat, but must be experienced close-up and on foot. The object of this chapter is to introduce the hiker to an area covering roughly 25 square miles (64 km²), all of which can be surveyed at the start from an elevation 656 ft (200 m) above sea level, perched in uplands dominated by igneous rocks. The shores along this part of the coastline continue to be shaped by passing storms of the kind survived by Cannon and his friends. But a far more intense interval of steady erosion occurred here near the conclusion of the Pliocene Warm Period about three million years ago (see the timescale in in figure 1.6). The early Pliocene Epoch is regarded as the closest analog to the accelerated warming we now experience in the earth's climate (Brierley et al. 2009). Basic factors that regulate our climate system involve the intensity of sunlight striking the earth's surface (always greatest around the equator), but foremost the concentration of atmospheric carbon dioxide (CO_2) now having reached a record level of 415 parts per million.[1] Carbon dioxide is a heat-trapping gas. As the Earth's atmosphere heats up, sea-surface temperatures (SSTs) also have the propensity to increase. The higher the SSTs during any given season, the higher the risk of hurricanes. It may seem highly theoretical, but Punta Ballena is one of those places on the Baja California peninsula where the aftermath of big storms sustained year after year throughout several hundred thousand years can be explored (Johnson et al. 2016).

FEATURE EVENT
Descending the Pliocene Punta Ballena Delta

The east end of Bahía San Rafael terminates at Punta Ballena (Whale Point), where uplands south of the bay are formed by tonalite, a dense igneous rock related to granite (similarly rich in silica) but often darker with a fine texture wherein it is difficult to see individual grains with the naked eye. Here, the rugged topography rises to a maximum height of 1,475 ft (450 m) above sea level. A dirt track that departs from the San Francisquito road brings the adventurous traveler to an extraordinary spot that amounts to a deep notch cut through an east-west ridge of solid tonalite (map 2, bold arrow points north). The walls on either side of the cut rise more than 160

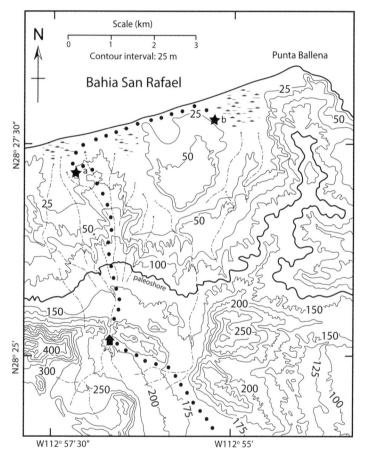

Map 2. Topography of the Punta Ballena area on the east side of Bahía San Rafael. The dotted line with a key lookout point (bold arrow) traces the featured hike for this locality. Dashed lines ending in small arrows mark the primary drainage pattern for the area. Two stars (labeled a and b) denote significant fossil localities near the coast. The former Pliocene paleoshore extends along the topographic datum at approximately 410 ft (125 m) above present-day sea level (bold line).

ft (~50 m), affording a kind of window with unexpected views both seaward and to a lesser degree behind. Not to be lightly dismissed, the topography directly south of the notch is a deeply incised bowl with a floor easily 1 square mile (2.6 km²) in area. The track leading through the notch and off the igneous promontory is solidly packed and runs north to the coast on a gentle incline for 3 miles (~4.75 km).

Spread out below, the view across the Punta Ballena basin has a mesmerizing quality, but details close at hand along the track are too important to overlook. Tempting as it is to descend to the bottom of the incline right away, a stop is called, and we pile out of the vehicle to examine sedimentary layers exposed adjacent to the track. Igneous rocks are behind us now, and here we find sedimentary layers of sandstone stacked to a height of 18 ft (5.5 m), with strange-looking vertical pipes that in places penetrate several feet (>1 m) of rock (figure 3.1). The sandstone is red, due to oxidation of the mafic (iron-rich) minerals eroded from the parent igneous rock. But the principal component is silica sand, and the sandstone layers are accordingly fine grained. Those vertical structures so prominent in cross section are fossil rhizomes, or root casts, from plants with long shafts buried in the crumbly rock. A whack with the rock hammer dispels any thought that the rock is solidly cemented. An eye to the surrounding ecology is always useful in solving questions about a former environment. Above, small acacia trees can be seen growing from the top of the cliff. Individual trees are evenly distanced from one another, but the tangle of roots from any single tree (both subhorizontal and vertical) is readily visible. The same kind of trees probably grew here in the recent past, and the more vertical structures represent burial of slender tree trunks in the accumulating soil. The bedded soil itself represents a pattern of repeated terrestrial outwash from the basin behind us within the massif. In essence, the sandstone deposit outside the cleft in the igneous escarpment represents an alluvial fan with a telltale fan shape that radiates outward from the opening.

Resuming the journey, our path descends on a slope with a 4 percent grade (equal to a 2.3° dip). Lines of topography (map 2) are splayed out about a mile (1.6 km) down slope, where ravines are entrenched on either side. Halfway to the sea, the ravines widen out, so that the trackway is more like a narrow ramp. The slope is

Figure 3.1. Track-side cliff exposure of silica-rich sandstone 18 ft (5.5 m) in thickness penetrated by root casts not unlike the root structures of living acacia trees growing at the surface. Photo by author.

readily measured by using the clinometer on a compass to gauge the dip by holding it edgewise at an arm's length to mimic the receding top layers of sandstone on either side to the east and west. The halt for this exercise also affords the opportunity to check the weathered surface of the sandstone for fossils. Disarticulated valves of fossil pectens (*Aequipecten abeitis* and *Pecten vogedesi*) are scattered about. The sandstone downslope has changed color, no longer red but beige in tone. We have traversed the transition from an alluvial fan to a distinctly marine setting. The erosion that eats through gullies in the landward direction reveals alternating layers of siltstone and sandstone and even a conglomerate layer composed of tonalite gravel. Ever downward, the worn trackway edges closer to one of the ravines on our western flank.

At the bottom of the ramp, the track turns to the west and rounds a corner where limestone is encountered for the first time. This is no ordinary limestone, because it is composed almost exclusively of fossil sand dollars (*Dendraster granti*) with a few fossil pecten shells mixed in. The first outcrop (figure 3.2) is only 30 in (75 m) thick, but the concentration of sand dollars is extraordinary and perhaps unique as a fossil deposit to this particular place. Elsewhere around Baja California, fossil deposits formed entirely of corals, coralline red algae (rhodoliths), and oysters are commonly found in Pliocene and Pleistocene strata (López-Pérez and Budd 2009; Johnson et al. 2009a; Johnson et al. 2009b). In 30 years of exploration along the gulf coast, I have not encountered anything else like the mix of sand dollars found here.

Exploring inland along the ravine, the full thickness of the sand-dollar limestone increases to 7.5 ft (2.3 m). Many of the sand dollars are unbroken with a diameter typically 4 in (10 cm) across. These represent the adult growth stage. A fast survey through the profile exposed at the toe of the ramp reveals that while all the sand dollars

Figure 3.2. Limestone deposit dominated by fossil sand dollars (*Dendraster granti*), many of them complete and nearly all tilted in the same position out of the horizontal. Pocket knife for scale. Photo by author.

are tilted out of the horizontal parallel to one another, roughly 80 percent are oriented with the mouth apparatus pointed downward. By comparison, living sand dollars from southern California in the same genus (*D. excentricus*) are known by marine biologists to live in dense populations partially buried in sand with their tests uniformly tilted at a high angle (Merrill and Hobson 1970). Where gregarious, the offshore limit of the California look-alike occurs at water depths from 20 to 40 ft (~6 to 12 m), but the extreme limit of live individuals has been recorded at a depth of 180 ft (55 m).

Individual layers of limestone exposed a bit higher in the ravine are not nearly as thick as the amalgamated deposit at the toe of the slope. They show sand dollars preserved in much the same orientation within discrete packages more likely to represent a final resting place in a natural life orientation. In contrast, other layers incorporate a hash of broken adult sand dollars mixed with whole but tiny sand dollars as small as three-eighths of an inch (5 mm) in diameter. These appear to have been swept together in a turbulent flow of water to form a postmortem coquina. The general configuration of features in the lower part of the ramp not far above present sea level conforms to a massive delta system linked to the terrestrial alluvial fan up slope, all of which emanates from the distinct notch in the igneous escarpment at a present elevation some 575 ft (175 m) above sea level.

The combined structures, with casts of fossil roots and buried tree trunks within the alluvial fan and abundant fossil sand dollars caught trapped in the marine delta, are all of one ecologic piece, now hoisted above sea level by tectonic uplift and exposed in three dimensions by the deep ravines cutting landward in more recent times. The original paleoshore falls at the junction between these two components most closely parallel to the line of topography about 410 ft (125 m) above present sea level (map 2). Followed laterally to the east and west, this particular line approaches more closely to the steeper terrain of the igneous escarpment, but traces an outwardly bowed shape similar in size and shape to the frontal arc of the modern Playa Costilla delta in context with its point of origin from Heme Canyon north near Puertecitos (review chapter 2). How much older is the Punta Ballena delta than its modern equivalent? The answer is based on the age relationship of the extinct fossil sand dollar (*D. granti*) compared with other fossils found in strata both above and below.[2] The assignment falls within the middle Pliocene. Together with absolute age determinations from volcanic ash deposits elsewhere bracketing strata with those other fossils, an age estimate is derived for the origin of the Punta Ballena delta around 3 million years ago (Johnson et al. 2017).

More than a ravine, the low ground on the west side of the ancient delta extends inland for a mile (1.6 km) before it gradually attains an elevation of 80 ft (~25 m) above present sea level (map 2). It can be assumed that the former delta occupied at least part of this ground prior to subsequent erosion during the Pleistocene. It is instructive to explore the incursion on foot, because from the valley floor it is possible to gauge the thickness of accumulated delta sediments exposed in the valley wall. In fact, the thickness midramp amounts to 150 ft (~45 m), dominated by massive deposits of fine sandstone generally without fossils (Johnson et al. 2017). The modern Playa Costilla delta south of Puertecitos is barely dissected by entrenched distributary channels, and there is no way to know just how thick that structure might be. Returning to the vehicle at the toe of the Pliocene Punta Ballena structure, it is a good spot to take a lunch break while contemplating the buildup of the delta through successive waves of sandy sediment flushed out from the igneous landscape above.

The ground along the shore at the foot of the former delta is muddy. Although tire tracks continue east along the edge of Bahia San Rafael toward Punta Ballena, the decision is made to leave the vehicle where it stands and explore the coast on foot for the rest of the day. The going is easy over level ground, but we soon pass places where the tire tracks are sunk well into the mud. Three-quarters of a mile (1.2 km) on, we reach the midpoint of another valley parallel to the Pliocene delta where it intersects the coast. This one (map 2) is as wide as the valley on the west side of the structure, but extends farther inland before reaching a defining elevation of 80 ft (~25 m) above sea level. The surface is wet and dark near the shore, but turns into a dry and barren mudflat inland. Our shoes kick up dust from the clay as we cross slightly inboard to avoid the mud. Given the sharp rise of the Pliocene delta structure on the west side of the second valley, it is likely that it formerly occupied at least part of the ground evacuated by the valley. In contrast, the parallel east wall of the same valley is fronted by a structure of lesser height. How much larger might the Pliocene delta have been before the two valleys on its west and east flanks were incised? How and when were these valleys eroded? A wider picture of landscape development is required, and we continue eastward for another mile (1.6 km), crossing in front of an isolated feature rising to a maximum height of 165 ft (~50 m) above sea level.

The tracks turn slightly south a bit at the far end of the structure (map 2) and terminate. Here, the largest mudflat of all can be viewed isolated from the sea by low-lying sand dunes. There is no standing water on the flats, but the surface is dark, muddy, and entirely barren of any signs of life. Unlike the salt lagoon at the side of Volcán Prieto near Puertecitos to the north (see chapter 2), this lagoon is not crusted by salt. The dunes that close off the flats from access to the bay are stabilized by vegetation. It appears that little or no seepage of seawater has leaked through from one side to the other. With no possibility of exploring the valley from the wet mudflats, we push farther inland along the shoulder of the hillside and climb to gain elevation. Here, the exposed clay layers rise some 30 ft (9 m) above the surface of the flats below.

Layered deposits at this locality clearly are different from the Pliocene deltaic beds, not least of all due to the obvious absence of the unique sand-dollar limestone. Aside from the unusual limestone, most of the deltaic beds are formed by sandstone with only a minor part of clay. For the most part, the profile lacks any sign of fossils except for horizons separated by thick intervals of barren clay. The clay has yet to turn into hard stone, and it is easy to dig into it with the pick end of a rock hammer. The first and lowest level features a line of fossil oysters (*Ostrea palmula*) that grew in place as a single generation in a coextensive, sheet-like cover. All exhibit fully articulated shells (both valves present and attached to one another). At little higher in the succession, another level with fossil shells is present and represented by one of the Venus shells (*Chione californiensis*). As with the oyster shells, found below, these Venus shells constitute one horizon of fully articulated shells (figure 3.3) representing a single

Figure 3.3. External and internal molds of Pleistocene bivalves (*Chione californiensis*) in life position at the east end of Bahía San Rafael (see map 2, locality b marked by star). Pocket knife for scale. Photo by author.

generation that matured in place as a coextensive population seeded at the same time. The original shells are gone, and all that remains is an outer mold with a faithful imprint of growth lines and an internal mold that replicates the smooth inner surface of the shell. As encountered previously at the side of the salt lagoon close to Volcán Prieto (see chapter 2), the fossil Venus shell is the same and it remains extant as a species today in the Gulf of California.

A third level higher in the sequence is distinct for its fossil rhizomes preserved as vertical root casts 8 in (20 cm) in height (figure 3.4). These are different from the much larger rhizomes encountered on the alluvial fan earlier in the day, and related to small acacia trees. Their plant affiliation unknown, the rhizomes from the clay bank are smaller in diameter and more closely spaced together. Like the shell layers below, this horizon signifies a single spurt of growth

Figure 3.4. Fossil rhizomes preserved as vertical root casts at the east end of Bahia San Rafael (see map 2, locality b marked by star). Pocket knife for scale. Photo by author.

that overcame prevailing conditions to colonize an otherwise barren environment. Clearly, the marine fossils indicate brief episodes during which normal seawater flooded onto a mudflat. In posture, the rhizomes appear somewhat like pneumatophores belonging to the black mangrove (*Avicennia germinans*), but are too large. They likely equate with another salt-tolerant plant.

The day's excursion has come to an end with insights on past environments that swung between drought and times of repeated deluge with floodwaters spilling out from a sizable basin semisurrounded by an upland igneous topography to feed a massive delta during the Pliocene Warm Period. Later during the Pleistocene, the sides of the great delta were eroded in concert with fluctuating sea levels and the advent of eroded valleys. Contrasts juxtaposed between the Pliocene and Pleistocene provide food for thought on the return march to the vehicle and the drive back up the deltaic ramp and alluvial fan to the notch in the igneous escarpment.

SHAPE-SHIFTING IN A LANDSCAPE

What did the landscape around Punta Ballena look like before the Pliocene Warm Period? How was it incrementally altered during that interval of increased storminess and rainfall? And how did the landscape change, yet again, during the ensuing Pleistocene to appear as it does today? Not quite the slight-of-hand trick performed by a stage magician, but the answers to these questions entail some aspects like that of a truncated body when the magician's pretty assistant is sawed in half. For the geomorphologist, the challenge is a mass-balance problem in which the sum of parts removed from one place need to add up to the parts reembodied in another place. Unlike the assistant's dismembered body that is shown to be magically reconstituted, a sundered landscape cannot be restored to its original shape.

The exercise has as its starting point the establishment of the former Pliocene shoreline (Johnson et al. 2017). As observed during our excursion, the junction between the alluvial fan and marine delta sets the marker for this boundary at Pliocene sea level. Topographic lines above the marker are relabeled in positive increments of 82 ft (25 m), whereas the bathymetry below is shown by dashed lines in negative stages through equal amounts (figure 3.5). In effect, the present-day landscape is made to undergo subsidence in reverse. Two key adjustments are next programmed working from the area's present topography. The most critical is removal of the mudflats in the three valleys with parallel axes perpendicular to the modern shore. These are post-Pliocene artifacts of erosion eliminated by cutting and reconnecting lines of topography to cross the valleys with a renewed east-west orientation. For example, it may be estimated that the volume of Pliocene strata reinstated through this operation amounts to 131 million cubic yards (100 million m³). The other adjustment is to add a small amount of elevation to the various saddles along the igneous ridgeline in order to compensate for their post-Pliocene erosion.

In preparation for the mass-balance experiment, a grid delineated in block units 820 ft (250 m) on each of four sides is superimposed over the reconstituted Pliocene topography, half of which is below sea level (figure 3.5). With this implement in place, it becomes possible to calculate the volume of material eroded to excavate the distinctly

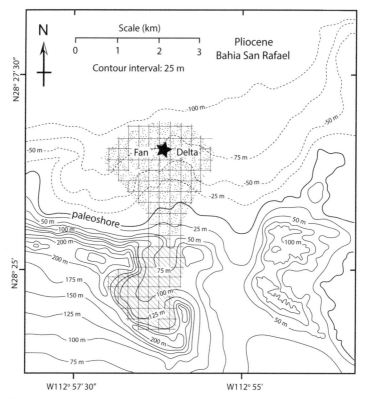

Figure 3.5. Topographic reconstruction of the east end of Bahía San Rafael for the middle Pliocene approximately 3 million years ago in which the Punta Ballena fan delta is restored to its original size, taking into account the material excavated by erosion from the adjacent terrestrial basin. Grid superimposed on the landscape is drawn in units 820 ft (250 m) on a side. Bold star marks the location of the sand-dollar coquina.

bowl-shaped depression in the igneous mainland as compared to the volume of sediments transferred to the Pliocene fan delta. To start with, the surface area of the delta in a bird's-eye view amounts to 1.3 square miles (3.5 km²), whereas the depression on the opposite side of the notch to the south registers as 1 square mile (2.6 km²). Clearly the delta covers a larger area, but this fails to take into account delta morphology, in which the outer arc of the fan thins out to a vanishing thickness but increases substantially toward the source. Likewise, this initial comparison overlooks the fact that the topography enclosing the bowl is abrupt. It means that, internally, the bowl is steep-sided in contrast to the delta.

The appraised volume of the two features may be calculated in vertical layers, adding block onto block in a three-dimensional construction constrained by each line of topography or bathymetry, in the case of the delta. The results yield a first-order estimate of 18.8 million cubic yards (~14.4 million m³) for the volume of sediments exported from the basin through the notch seaward to the delta. The volume of the Pliocene delta together with the adjacent alluvial fan is estimated as 16.26 million cubic yards (~12.4 million m³). The difference between the two amounts to 25 million cubic yards (~20 million m³). By this construct, it means that roughly 14 percent of the sediments carried away from the eroded basin cannot be accounted for in the calculated volume of the alluvial fan and delta. Given the grand scale of features under comparison, the difference is surprisingly small and might be accounted for by loss of the clay fraction of exported sediments suspended in seawater during storms and carried offshore by currents.

Certain mitigating circumstances must be accounted for in further contemplation of such a mass-balance exercise. The first is that the Pliocene alluvial fan and adjoining delta was steeper than the contemporary Playa Costilla delta near Puertecitos (see chapter 2). The latter drops 164 ft (50 m) from the mouth of Heme Canyon over a distance of 1.5 miles (2.4 km) to the sea's edge, whereas the Pliocene delta dropped the same amount over two-thirds of a mile (~1 km). It means that eroded sediments flushed onto the Pliocene delta system with each passing storm were likely to have been more turbulent descending across a steeper gradient. The second is that the modern watershed shown by the bowl-shaped depression behind the notch in the igneous escarpment is appreciably smaller than that for Heme Canyon to the north by at least an order of magnitude. It means that the tonalite rocks landward from Bahía San Rafael have been subject to an extreme degree of erosion compared to the ignimbrites in Heme Canyon. The third factor apparent from reconstruction of the Pliocene delta (figure 3.5) is that the sand-dollar coquina is located far downslope at what was likely to have been a water depth around 200 ft (~60 m), which exceeds the maximum depth for a closely similar species of sand dollar (*D. excentricus*). The broken tests of adult sand dollars that dominate the coquina lend a sense of extreme energy delivered with violent floods during the Pliocene Warm Period.

CONTRAST WITH THE ENSENADA BLANCA BASIN

A published satellite image dating from 2003 covers the adjoining areas of Punta Ballena and the Ensenada Blanca basin at San Francisquito to the east in which igneous tonalite and granite uplands are well differentiated through false-color analysis.[3] The same field of view is shown under natural illumination in an aerial photo taken in 2016 during a commercial flight between Los Angeles and Loreto (plate 3). A hike of about 7 miles (11.25 km) around the perimeter of the Ensenada Blanca basin is extolled in an earlier guidebook.[4] Pliocene strata within the confines of the 4-square-mile (10 km²) basin include conglomerate, limestone, and sandstone together with abundant fossil bivalves and brachiopods. A small river delta enters the basin from the northwest, but its size is miniscule compared to the neighboring Punta Ballena delta system. During the later Pliocene time, access to the Ensenada Blanca embayment was through a pair of openings from the Gulf of California, but they were relatively restricted. What is most striking about the former basin is that it would have made a safe harbor during the kind of hurricane-strength storm weathered by Ray Cannon and his friends off Isla San Esteban in 1957. There were no human beings around at that time and, of course, no sea-going vessels. The reason why the basin never became choked with deltaic sediments was due to the low topography of the surrounding granite hills that rose protectively 260 ft (~80 m) above the inlets, but much less so only 65 ft (20 m) around the back margins. The igneous ramparts fronting Pliocene Bahia San Rafael were not only equal in stature to the sentinels guarding the entrance to the neighboring basin, but featured a distinctive notch through which a mighty alluvial fan and marine delta immerged. Both features remain part of the Baja California landscape inherited from the past, still readily visible today.

4

The San Basilio Embayment and Pliocene Volcanic Islets

In San Basilio, I only want to hear
what the Earth is drumming into me.

—*Sylvia Holland, "Learning This Place"*
ESSAY IN *Reflections by the Sea* (2016)

AS A PLACE name, San Basilio is scarcely mentioned at all in the extensive travel literature on the Baja California peninsula and its gulf shores. John Steinbeck makes no mention of stopping at Ensenada San Basilio in *The Log from the Sea of Cortez*. Joseph Wood Krutch, a veteran traveler with a poetic eye for scenery, never got to San Basilio. Among those who lay claim to walking the entire coastline or kayaking the full length of the Gulf of California from the Colorado River delta to Cabo San Lucas,[1] not a single word of recollection regarding San Basilio is committed to published accounts. Without hesitation on my part, San Basilio Bay is surely among the most pristine and tranquil of coastal settings found anywhere around the entire Gulf of California. Quite simply, it is one of the best secrets hidden by the land. There are only two ways to arrive at San Basilio (see figure 1.7). By boat, the trip from Loreto northward along the coast amounts to a sea journey of 19 miles (~30 km). By car or truck via ranch roads east off Mexico Highway 1 that converge on the bay, the distance from Loreto is around 36 miles (~58 km), depending on available options.

My initial exposure to the bay at San Basilio occurred on a bright January day in 2001, when local fishermen were hired to bring two boatloads of geology students and their instructors on a day trip north along the shore from Punta El Mangle to Punta Mercenarios at the southeast corner of the bay and back again. It was a round trip of barely 10 miles (16 km), but one of the most memorable from my many years of fieldwork in Baja California. Our studies at El Mangle stretched out over three field seasons and led both to technical and popular accounts of a geologically fascinating place.[2] By comparison with any other place I've studied, San Basilio turned out to be the most complex and physically demanding of puzzles to interpret in regard to its geologic origins. On subsequent visits, I used San Basilio as a kind of final exam for students who were asked to complete mapping exercises. Only gradually did the place begin to reveal its secrets, but the story written in rocks came into better focus after a hard push in March 2016 that resulted in a published summary of Pliocene events (Johnson et al. 2019a). Additional efforts to study a Holocene storm deposit at nearby Ensenada Almeja also came to fruition (Johnson et al. 2019b).

During that first visit in 2001, time allotted for exploration was limited to a single afternoon. I recall scanning with binoculars the sea cliffs and sandy shores lining the bay through an arc of 3.75 miles (6 km) from end to end. That's when I spotted the home so artfully tucked into the cliffs across on the far side of the bay. The buildings were painted a desert-tan color that made them meld into the cliff side. No roads were visible from our vantage point. I wondered, who might possibly live there? Who would have the means to build and maintain such a house isolated in such a scenic spot? Details were filled in later on. The place was designed by a renowned European architect for a wealthy client, a Spanish contessa, who first sailed into the bay sometime in the mid-1980s.[3] She named the place *Agua Magica* (Magic Water). I can only imagine how substantial construction materials had to be brought up the coast from Loreto by barge. Off-grid in a spot so remote from freshwater springs, the domicile was initially fitted with its own saltwater desalination plant.

The terrace in front of the house included a pool, the edge of which closed on infinity with the bay's waters below. The very basis of sustainable life: fresh water in a parched desert landscape, became an extravagant showpiece. It is a conundrum I have often confronted in my travels. Is the natural beauty of a place somehow violated by the desire to understand the forces by which it was made? I think not, but the answers are not always easy to find. Perhaps the human species is driven to ask such questions partly in search for itself. Imperfect as they may be, the excursions laid out in the following rambles require three days for onsite reflection around and about San Basilio.

FEATURE EVENTS
Three Rambles Across Pliocene and Holocene Timelines

As rendered (map 3), the topography around Ensenada San Basilio, including the jewel-like Ensenada Almeja to the north, covers a mere 4 square miles (~10.25 km^2). The complicated geology in the larger surrounding region has attracted little attention, but oddly enough a team of Italian geologists produced a report on the adjoining Cerro Mercenarios volcanic center (Bigioggero et al. 1995) that adopts the same name attached by mariners long ago to Punta Mercenarios at the bay's southeast corner. On approach to Loreto International Airport from the north, incoming flights descend west of the extinct stratovolcano to offer a magnificent view of the edifice having a maximum elevation of 2,600 ft (~790 m) above sea level and covering 58 square miles (150 km^2) of rugged terrain.[4] It is a district untouched by human progress, in which wildlife including deer and mountain lions freely roam. Geologically, Miocene lava flows formed by andesite dominate more than 80 percent of the complex, and give the radial structure its characteristic red overtone. In contrast, the principal volcanic rocks enclosing Ensenada San Basilio are formed by dark rhyolite, which chemically is identical to granite with abundant silica (SiO_2), but with tiny mineral grains due to rapid surface cooling.

The earlier effort by the Italians focused on the igneous rocks, giving perfunctory attention to the associated sedimentary rocks at the heart of this story. Visitors who find their way to Ensenada San

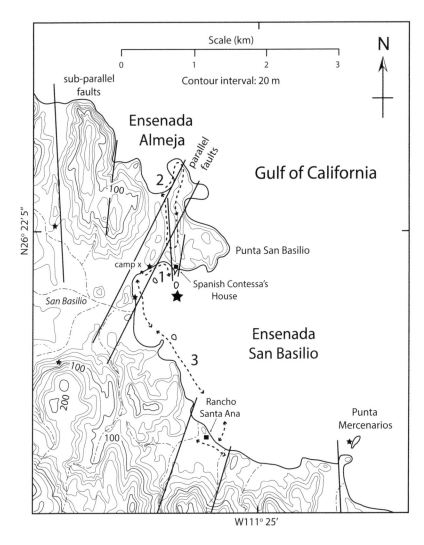

Map 3. Topography around Basilio and Almeja bays off the north flank of the Cerro Mercenarios Volcanic Complex. Long and short dashed lines mark the central drainage for this region. Bold dashed lines with double-headed arrows trace the featured routes for three rambles through parts of the district. The large star near the center of the map marks the premier locality, with a Pliocene volcanic islet and associated sedimentary and fossil deposits. Small stars mark additional sites at the center of Pliocene island volcanos.

Basilio by boat, anchor in the bay. Those who arrive overland camp on the beach at the northwest side of the bay (map 3). The following annotated rambles advance from a short hike across part of the north shore on the first day, to a longer hike around Ensenada Almeja and up over the terrain on its east flank during the second day, and conclude with a trip by kayak to Rancho Santa Ana and back to camp on the last day.

First Day (visit to the cliffs below the Spanish contessa's house): A preview of the ground covered by the introductory ramble is provided by the view at low tide over the "Birthday Cake" sea stack and beach leading from the camp site, including Cerro Mercenarios on the distant skyline (figure 4.1). It is a comfortable stroll from camp along the beach with the possibility to explore the shores around the "Birthday Cake" with its cardón cactus candles (*Pachycereus pringlei*). A swimsuit and water-resistant footwear make a good choice for this hike. A wide-brim hat and long-sleeve shirt for protection from the intense sun are advised.

Figure 4.1. View along the northern and western shores of Ensenada San Basilio with the "Birthday Cake" sea stack (center left) connected to the shore at low tide. Note sailboat masts for scale. Photo by author.

Only a short distance east from camp, the first cliffs rise up behind
the beach to project the uneven look of vertically fractured rocks.
Here we encounter the igneous rhyolite in but one form among other
variations. Arriving as I first did from the sea, these rocks express
the same morphology as the jagged spires of rhyolite in sea stacks
at Punta Mercenarios. Exploring a bit farther to the eastern limit of
these rocks, it can be seen that limestone crowded with fossil pectens
forms a cover draped against the side of the igneous structure. The
direct contact between the sedimentary and igneous rocks, however,
is not well exposed. Clambering through the talus pile of rough rhyolite stones, a kitchen midden with modern pecten and oyster shells
is next encountered within a stone's throw of the abutting limestone.
Taking time to search among the stony debris, it is possible to pick
up flakes of obsidian that were worked by the indigenous people who
made this place a favored camping spot. The flakes are the residue of
cutting tools fashioned by these earlier inhabitants, and we will discover where they obtained the black, glassy rocks during tomorrow's
ramble. For the moment, it is enough to contemplate the nomadic
life style of San Basilio's early visitors and how they drew daily sus-
tenance from their surroundings.

Following the falling tide, as the indigenous people clearly did
from this very beach, it is possible to cross to the "Birthday Cake"
while wading at first in water barely ankle deep. Harvesting clams
from the exposed flats around the north side of the sea stack was
part of life's routine in a setting that provided ample resources. The
cliffs of the sea stack are not climbable, vertical and as high as 50 ft
(15.25 m). But beneath the cliffs on the west side, one can sort among
fallen limestone blocks with abundant fossil pectens that originated
from stratified layers at the irregular top of the sea stack. The rocks
below the limestone are igneous, but unlike the fractured rhyolite
across the way near the beach. In contrast, they are massive and
smoothed by the waves. Here and there, a faint pattern of shallow
dimples about a foot (30 cm) in diameter is evident in the dark rocks
assumed to be another variation on rhyolite. Two observations are
possible. The first is that the "Birthday Cake" has no obvious link
with the rocks on the opposite side of the channel where we came
from, although the rocky ground exposed by the tides indicates

significant reduction in that direction (figure 4.1). Following from the first, the second observation is that the contact with overlying limestone is well defined but represents a top cover rather than a side drape as surmised from the rocks along the beach. Physical differences in features relatively nearby one another might be explained by fault lines that separate the different forms of igneous rocks and their contrasting sedimentary rocks. If so, such faults would represent planes along which movement (up or down, sideways, or even a combination of the two) brought things closer together that may have formed at different times or in somewhat different places. It is something to ponder as we cross back to the beach and continue eastward.

A big acacia tree farther along the beach is festooned with small flags and various wooden tokens like an ornamented Christmas tree. The colorful regalia are emblematic of the custom whereby yachtsmen leave behind some sign of their visit. Not far beyond the tree, a path departs inland from the beach and climbs in elevation through the hills. We start up the path to reach a better lookout over the bay, which improves with altitude farther up the ravine's west shoulder to a spot where the rocks below the Spanish contessa's house can be viewed to advantage (figure 4.2). The geologic features are complex and they are the chief object of the day's ramble (map 3, marked by a large star). Uncovering the mystery of these rocks as a distinct variant of rhyolite is the ultimate goal of the day's excursion.

Flat-lying strata push seaward away from the coast in what can be described as a narrow sedimentary bridge. At the end of the bridge, layers turn upward in a ramp-like configuration. The top of the ramp has broken off and fallen to the shore below as an enormous block. Farther seaward from the ramp, isolated spires of rhyolite, much like the igneous rocks at the beach, point upward to the sky. In the distance, the same rugged spires mark the southeast corner of the bay at Punta Mercenarios. The scene is iconic to the San Basilio basin. If the relationships among these features can be worked out, we will have solved much of the mystery beneath the magic waters of San Basilio. The principal question is whether or not the bend in strata resulted from some cataclysmic tectonic force after the layers were first deposited, or formed individually one layer at a time during their

Figure 4.2 View from an elevated lookout point to the southeast across Ensenada San Basilio, with the rocks below the Spanish contessa's house in the foreground and the Punta Mercenarios sea stacks in the far distance (zodiac boats for scale). Photo by author.

deposition in a naturally recumbent position. The answer demands a closer examination of the entire structure. That task promises to fill the rest of the day, but we must get there to make the inspection. The cobbled beach brings us only so far (figure 4.2) before it is necessary to wade through knee-deep water in order to reach the upturned strata at the side of a tiny beach.

Data summarized in the graphic profile for Pliocene strata exposed at this key locality (figure 4.3) were originally compiled bed-by-bed during a visit in March 2017. At the base of the section accessible at low tide, there occurs a wedge of coarse volcanic breccia with rough blocks of rhyolite changing upward in size through units 1 and 2 into a mixture of worn rhyolitic cobbles and pebbles set in a matrix of volcanic ash.[5] Measured perpendicular to bedding, these units exceed 30 ft (~10 m) in thickness. The junction between overlying limestone and the inclined surface of the volcanic conglomerate draws a sharp break, or unconformity, between

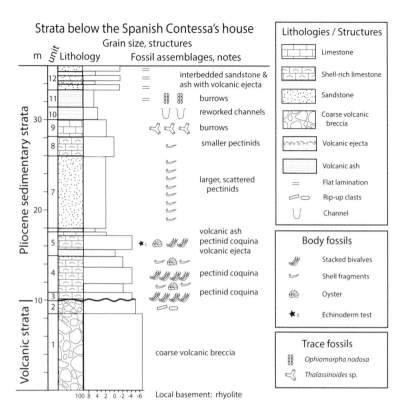

Figure 4.3. Coastal succession of key strata directly south of the Spanish contessa's house with total thickness of 116 ft (35.35 m). Coordinates: 25° 22' 13.40"; 111° 25' 44.93 W).

the two rock types. A similar boundary was encountered earlier at the "Birthday Cake," but it was not possible to view it close-up, exposed as it was high in the cliff face. Here, the junction can be examined minutely at nose length, and it constitutes an abrupt change in rock types.

Directly above the break starting with unit 3 and thereafter in the lower parts of unit 4 (figure 4.4), disarticulated pecten shells are pressed together in a dense mass that reflect a postmortem accumulation with valves stacked congruently within one another. In cross section, roughly half of the shells are turned to stone with the outer convex-shell surface pointed downward. The other half show the outer convex surface pointed upward. The mixture is chaotic in the context of recurrent limestone layers typically 8 in (20 cm) thick. The pectens did not live exactly here, but were swept downslope in a turbulent flow from above. This interpretation is reinforced by the presence of mud balls showing a concentration of shells pressed together in a semispherical shape, but what in cross section has the appearance of a pinwheel with individual valves radiating outward. Mud balls with incorporated shells in the same pattern are reported from Pliocene deposits found father south in the Gulf of California on Isla Monserrat.[6] They are interpreted as part of a gravity slide, where sticky mud swept downslope, picking up loose pecten shells in a rolling mass during an underwater avalanche. The correlation makes a strong argument that limestone layers at this locality only appear to be bent, but instead were deposited in a naturally inclined state. Thus, the repetitious pecten coquinas are part of a sedimentary ramp that grew layer by layer with additions of shells transferred downslope from above. We find these layers, now, in their final resting place where successive slides came to a full stop.

Figure 4.4. Densely packed coquina of fossil pecten shells near to the base of Pliocene strata below the Spanish contessa's house (handle on rock hammer for scale is 6 in, or 15 cm, in length). Photo from D. H. Backus.

Belonging to unit 5 (figure 4.4), the next interval amounts to 8.5 ft (2.5 m) of limestone formed by a similar shell coquina, but also incorporating blocks of welded volcanic ash as large as 16 in (~40 cm) in diameter. Greater or smaller, these have every appearance of having been dropped like bombs from above. Volcanic ejecta is the appropriate name for such objects with a volcanic origin that are entirely extraneous to the surrounding rock composition. At the top of this unit, an extensive bedding plane is exposed to reveal large pecten and oyster shells (disarticulated) together with bits of broken sea urchins (*Clypeaster* sp.). What comes next, however, gives us pause to reflect on the progression of events. Unit 6 is a layer of volcanic ash only a foot (30 cm) thick, but sufficiently soft in consistency compared to the resistant limestone below to result in a deep erosional recess in the profile (figure 4.3). The early afternoon sun casts a shadow below a sandstone overhang. It makes a suitable spot for a lunch break, sitting in the shade with feet pointed downslope propped against the ash layer.

Where, exactly, do we find ourselves in the context of long-past events that followed one after another on a former seafloor in what today is a large, open bay called Ensenada San Basilio? We have worked our way across time, halfway upward through a pile of sedimentary layers that accrued above a thick wedge of volcanic-derived breccia and conglomerate. Frankly, everything we've seen of the limestone testifies to an effusive accumulation of dead shells. We can conclude that none of these mollusks lived precisely here, because none are preserved intact with united valves in life position. We sit in a dead zone surrounded by a copious volume of shells, while chunks of volcanic rock descend now and then like cluster bombs through the water column from above. But now, here at our feet where a recess in the cliff is eroded (figure 4.5), we experience the remnants of an explosion that sent volcanic ash streaming down through the water like a curtain that surely blocked out any sunlight to otherwise reach these depths. Do we feel a tremor in the seafloor on which we sit? The violence of an explosion sufficient to send a great volume of volcanic ash into the air surely caused the ground to shake. Just how close are we to the volcano from which the larger ejecta and all the ash emanated?

Figure 4.5. Strata below the Spanish contessa's house with emphasis on the erosional recession in the dominant limestone due to a bed of soft volcanic ash located 22 ft (6.75 m) above the base of the Pliocene succession. Photo by author.

The volcano is close by, but presently out of sight to the south, with its original volcanic orifice not much farther away than the length of a soccer pitch. The jagged spires of fractured rhyolite are manifestly evident as the clogged throat of a small volcanic edifice. The term for such a remnant is a volcanic neck. The top of the volcanic neck in front of the Spanish contessa's house rises abruptly some 30 ft (~9 m) above the surface of San Basilio Bay. How deep it plumbs below the present seafloor is anyone's guess. But the wedge-shaped unconformity now below us at sea level represents the original underwater slope that formed a three-dimensional skirt wrapped around the waist of a Pliocene volcanic islet.

Venturing back into the sunlight after our repast, the opportunity arises to climb toward the top of the sedimentary ramp with the rhyolite spires in the background. Here, we find more bits and pieces of fossil sea urchins (again, from a species in the genus *Clypeaster*). Crossing the gap with the eroded ash bed, we stretch our legs on the flattop sedimentary bridge below the contessa's house. These rocks are dominated by fine-grained sandstone bearing scattered lenses of small, disarticulated pecten shells. The overall thickness of unit 7 in the succession (figure 4.3) amounts to a whopping 26 ft (~8 m). Impossible to miss, a shallow basin covered with salt crystals sits astride the midpoint of the bridge. The salt is the product of evaporation from a pool of seawater that splashed onto the bridge during the last big storm. Normal seawater contains 35 parts per thousand of different salts, the dominant one being sodium chloride (NaCl). When approximately nine-tenths of the volume of seawater disappears through evaporation, common table salt begins to crystalize.

The last major storm to impact Ensenada San Basilio was Hurricane Odile on September 15, 2014. A guest at the Spanish contessa's house who stayed to brave the hurricane made a video clip that documents the storm's fury. As reported by Muría-Vila et al. (2018), the winds south at Loreto were clocked at 90 mph (~113 km/hr). Out at San Basilio, wind-driven waves dashed against the rhyolite spires in front of the Spanish contessa's house and spray overtopped the crest of the sedimentary ramp.[7] Anyone foolish enough to venture down to the sedimentary bridge from the shelter of the house during the storm would have been swept away and battered against the rocky shore. The benign after-mark of such a storm is the pool of seawater left standing on the bridge, leaving a telltale residue of sea salt once the sun returned for some days at a stretch.

Continuing with our account of the upper strata preserved on the sedimentary bridge, unit 8 (figure 4.3) amounts to a 7.3-ft (2.25-m) thickness of silty limestone bearing scattered pecten shells. An exposed bedding plane within this interval provides a suitable spot to obtain a firm reading on the local dip of strata that registers as a 12° dip to the northeast (figure 4.3). It means that the sedimentary slope on this flank of the Pliocene volcano was relatively steep. The overlying unit 9 continues with more silty limestone, but here with fewer shells and the first appearance of burrows from an arthropod similar to the extant ghost shrimp (*Neotrypaea californiensis*) of Alta California. When preserved, the burrows are regarded as trace fossils and given their own name in classification (*Thalassinoides* sp.). The succeeding units 10 through 12 entail layers composed of fine

sandstone interbedded with volcanic ash (figure 4.3). Body fossils
are scarce within these sandy intervals. These bottom sands appear
to have covered the seafloor around the volcano islet, burying all
below.

The afternoon has advanced, and we must return to camp before
our first day at San Basilio concludes. The tide has begun to advance.
Crossing below the towering strata of the sedimentary bridge
requires us to wade through water hip deep to access the beach
linked to camp where the day's ramble began. The return distance is
not far, only a third of a mile (0.5 km). Once on the beach, it is tempting to turn and look back at the sedimentary bridge that merges
with upturned layers tracing the underwater slope of a small volcano.
Knowing something about the power of hurricanes that episodically
enter the Gulf of California, it is astonishing to realize that the entire
outer perimeter of the contessa's volcanic islet has disappeared due
to marine erosion. That is to say, everything in the diagrammatic
restoration of the islet's seaward profile (figure 4.6) has now vanished. Indeed, all that remains of the edifice itself are the jagged
spires of the volcanic neck. Moreover, a sizable gap exists between
the spires and the upturned ramp of Pliocene strata. Retracing our
steps past the vertically fractured rhyolite cliffs encountered during
the morning's ramble along the shore, it is yet more astonishing to
realize we pass through the dissected heart of another volcanic islet
in Pliocene San Basilio.

Perhaps in our search for a model to explain the geologic phe-
nomena surrounding these magical waters, we demand too much
of a black-and-white answer. While I continue to believe that the
sedimentary ramp (figure 4.6) formed mostly in place, layer-on-layer
through the passage of Pliocene time, it is possible that the throb-
bing, magma-filled chamber beneath even a small volcano effected
some degree of uplift to increase the tilt of the sedimentary ramp.
Asleep in camp close to a beach on the enchanting Ensenada San
Basilio, dreams of silently smoking volcanic islets on its perimeter
drift into place, only to be darkened by descending storm clouds.
The morning's arrival brings a new day on the placid shores of San
Basilio that recalls no violence of Hurricane Odile or of volcanic
eruptions from the Pliocene past.

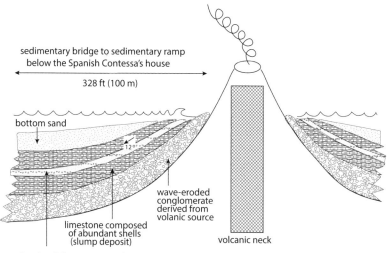

sedimentary bridge to sedimentary ramp
below the Spanish Contessa's house

328 ft (100 m)

bottom sand

12°

wave-eroded
conglomerate
derived from
volanic source

limestone composed
of abundant shells
(slump deposit)

volcanic neck

volcanic ash from eruption that interupts limestone deposition

Figure 4.6. Interpretive restoration showing the complete north-south profile of the Pliocene volcanic islet below the Spanish contessa's house. Parts beyond the volcanic neck to the south are now missing due to post-Pliocene wave erosion.

Second Day (visit to neighboring Ensenada Almeja): A preview of the ground to be covered is shown by the view at low tide across a coastal boulder deposit (CBD) that partially blocks the entrance to Ensenada Almeja (figure 4.7). The pristine beach at "Clam Bay" opens due north onto the Gulf of California (map 3). It is accessible either by road from camp or by way of the trail leaving Ensenada San Basilio from the acacia tree festooned with boaters' paraphernalia. Either way, the distance from camp to the circular loop starting at the beach is about three-fifths of a mile (1 km). Focus for the day's ramble is on the north-south valley linking Ensenada Almeja to Ensenada San Basilio and the parallel ridge to its east that terminates above the Spanish contessa's house. The valley is serviced by its own gravel road over a level surface, but the terrain along the ridge top is off-trail and requires a climb to reach a modest elevation. The sole acknowledgement of the San Basilio territory in the popular literature derives from the Spanish-language book *Oasis de Piedra*,[8] issued after the English-language version in 2008. Both feature beautiful photographs of rocks and animal life, but

Figure 4.7. View from sand dunes above the beach at Ensenada Almeja showing the peculiar coastal boulder deposit (CBD) that partially blocks the bay's opening to the north. Photo by author.

a stupendous aerial photograph of Puerto Almeja and its adjacent rocky ridge appears only in the Spanish edition. The day's mission is to acquire a geologist's intuition regarding the landscape in that single image by photographer Miguel Ángel de la Cueva.

The turquoise waters of "Clam Bay" are stirred by seasonal winds out of the north, especially during the months of January through March. But even during winter time, the winds are becalmed for days at a time. Aptly named, the bay is rich in mollusks that contribute their shattered shells to the carbonate content of the beach and contemporary dunes behind. Many if not most of the bivalves populating the bay, such as the chocolate shell (*Megapitaria squalida*), are infaunal in habit. It means they live their lives below the sandy surface of the bay's floor in burrows excavated by a fleshy foot capable of exerting powerful digging movements in coordination with the attached valves. A clam's life is brief on a human scale, rarely more than a half-dozen years. When a clam avoids harvesting by humans and dies a natural death of old age, the winter winds that ripple the bay's waters bring the empty shell to the surface where it tumbles in the waves and is abraded through contact with other shells. Over time, the ongoing breakdown of mollusk shells adds materially to the buildup of the beach. Moreover, the north winds blow some of that sand into the dunes behind the beach. The pattern at Ensenada Almeja is the same found on the north-facing shores of Ensenada Muerto west of Punta Chivato and along the north coast of Isla Monserrat, where coin-shaped disks of worn shells are commonly washed onto the beaches.[9]

A sketch map giving the layout of the Almeja boulder deposit is drawn to scale with a gridded overlay marked off in units 32.8 ft (10 m) on a side (figure 4.8). Cliffs rise vertically some 16.5 ft (5 m) in height as the east wall against which the deposit abuts. Another variation on igneous rocks called banded rhyolite form these cliffs. The name is especially apt, because thin bands, or layers with the thickness of laminated plywood, are readily apparent in cross section (figure 4.9). The hiker soon encounters these cliffs a short distance from the beach along the east side of "Clam Bay." Far from the look of a flat stack of plywood, the typical swirling pattern found here gives the impression of extreme plasticity in magma as it flowed out in discrete pulses from the ground. But the flow also was sticky, and perhaps a better analogy is the nature of warm taffy that is easily folded on itself before it cools and hardens. It is the third type of rhyolite encountered on our tour around San Basilio. Previously, the most distinctive form was represented by vertically fractured rhyolite eroded as spires attributed to volcanic necks. A subtle variation, as yet unexplained, was encountered at the "Birthday Cake" sea stack, where its walls are oddly dimpled.

Prior to reaching the first boulders adjacent to the cliffs, terraces cut in solid rhyolite are found buried by sandstone dunes entirely consistent in composition with the modern dunes behind the beach. These "fossil" dunes are Pleistocene in age, dating from approximately 125,000 years ago.[10] Such lithified dunes are commonly seen many places around the Gulf of California, such as at Punta San Antonio and Punta Chivato farther north, as well as the north shores

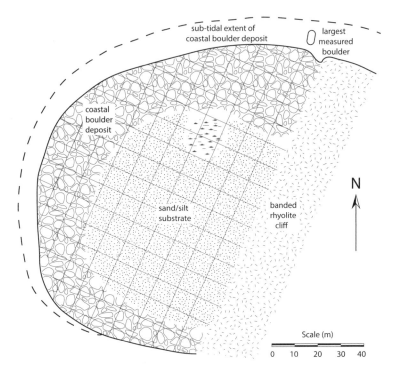

Figure 4.8. Map view of the semicircular boulder deposit bordered by cliffs of banded rhyolite at Ensenada Almeja (modified from M. E. Johnson, Guardado-France, E. M. Johnson, and Ledesma-Vázquez 2019).

of Isla del Carmen. We will encounter more of these lithified dunes ensconced in the north-south valley linking Ensenada Almeja and Ensenada San Basilio at the end of the day's ramble.

It is evident on arrival at the first line of boulders along the south margin of the deposit that all are derived from the same kind of banded rhyolite found in the adjacent cliff side. Consulting the gridded map (figure 4.8), the structure's subaerial exposure amounts to a total area of 15,550 square yards (~13,000 m²) of which the boulder field around the rim occupies half that amount. The arc of the boulder field is like a rampart that increases in height to the north, the seaward base of which extends well below the waterline. Crossing north within the enclosure, the ground becomes softer and merges with a small pond below an abrupt wall of boulders stacked to a

Figure 4.9. Close view of swirls in banded rhyolite exposed in the cliffs along the east side of Ensenada Almeja (pen for scale = 5.25 in, or 13.5 cm). Photo by author.

height of 6.5 ft (~2 m). The desiccated soil around the water hole is sun-cracked, but loose pieces pulled from the ground are held together by strands of organic matter. The scenario is reminiscent of the closed lagoon at the side of Volcán Prieto (see chapter 2), although no salt crust has precipitated. Here, the prevalent plant cover is the *Saladillo*, or salt bush (*Atriplex barclayana*), which is a salt-tolerant plant adapted to an environment with salty soils. Organic material infused in the dried clay is likely to include root structures from this plant. Although the nearby pool is shallow, its open mud-bottom hints at a substantial accumulation of sediment in which microbes like those from other closed lagoons could be present. The spot is ripe for further investigation by biologists interested in microbes.

Some of the largest boulders piled along the north end of the semi-circular structure are enormous, as much as 8 ft (~2.5 m) in diameter. Based on the density of banded rhyolite and the volume of a particular boulder, its weight can be estimated (Johnson et al. 2019b). The largest boulders amount to 4.8 tons (4.35 metric tons), and would require a tremendous force to shift into place. The Almeja structure is unconsolidated, meaning that the boulders in the deposit are loose. The fact that constituent boulders are not cemented together like conglomerate tells us that the deposit is geologically youthful. How could this giant, semicircular structure form? Indeed, might it still be in the process of continuing growth? Clues as to source are found at the end of the path leading north along the cliffs, where a narrow cove opens between rhyolite bedrock on the right hand and rhyolite boulders to the left. Exposed rhyolite is part of the headland pointed due north, where the natural state of weathered rocks is readily observed. Here, the headland is dissected by bedding planes parallel to the original flow of the banded rhyolite, as well as vertical fissures that intersect bedding planes at right angles. In effect, the headland is caught in the act of breaking apart into ready-made, boulder-size chunks that are cubic to rectilinear in shape. Production of rhyolite boulders results from hydraulic pressure exerted by wave surge against the natural joints and bedding planes in the sea cliffs. Blocks larger and smaller in size that sit on bedding planes are predisposed to slide downslope into the sea, assisted by the force of gravity once they are sufficiently loosened by storm waves.

This is not an everyday process, but one that awaits the arrival of a hurricane. Mathematical equations are available (Johnson et al. 2019b) through which the wave height generated by a major storm can be estimated having sufficient force to lift a block of given size free from rhyolite bedrock. The largest boulders in the Almeja deposit would require waves of prodigious size to make them budge. Roughly 30 percent of the boulders in the deposit are reckoned to exceed a short ton (2,000 lbs, equivalent to 907 kg). The wave height necessary to shift those boulders is close to 26 ft (~8 m). The winds of Hurricane Odile produced storm waves that crashed over the rock structures below the Spanish contessa's house at about that height on September 15, 2014. The same storm surely impacted the rhyolite headland alongside Ensenada Almeja. No one was foolish enough to be out and about to film such a spectacle, but many of the stones in the Almeja boulder deposit were sure to have shifted during that event.

Evidence is circumstantial, because there is no eyewitness account. However, additional lines of reasoning point to the work of hurricanes in the Gulf of California as capable of forming coastal boulder beds in incremental stages over the last few thousands of years within the realm of Holocene time. With this in mind, we set out on the return loop for the day's ramble, starting the climb up the Almeja headland to reach a lookout point (map 3). There is no actual trail to follow, but by tracking back and forth from one ledge to another, we arrive at a spot some 26 ft (~8 m) above sea level where a lunch pause may be taken while enjoying the view. With a nearly 180° view from east to north to west, the asymmetry of the Almeja boulder bed makes a strong impression. That is to say, the deposit is not symmetrical on both sides of the headland. It occurs only within Ensenada Almeja. The other factor we must keep in mind is that the structure is strictly semicircular in shape, with no trace of reworked boulders at the base of the adjacent cliffs against which the two ends abut. Also, the fact is that maximum boulder size gradually decreases from the northeast end next to the headland to the southeast end through a 1,000 ft (~305 m) curve. In other words, the largest rhyolite boulders occur to the northeast, but rhyolite cobbles and even pebbles are found among the giant boulders. At the opposite end of the deposit, there are no more boulders, only larger cobbles mixed with pebbles. It means that the arc of the complete deposit has its sole origin at the headland.

How do such observations square with the normal behavior of hurricanes that pass through the Gulf of California from south to north? As discussed regarding Hurricane Marty and its impact on Loreto on September 19, 2003 (see chapter 1), forward movement in the northern hemisphere is accompanied by a swirling counterclockwise movement with the storm's highest winds in the "right-front quadrant." The energy spins off rain bands that move in an east to west direction, while also pushing large waves in the same direction. It means that a storm like Hurricane Odile would push waves into Ensenada San Basilio from east to west, while also striking the headland at Ensenada Almeja with greatest force on the east side, crossing the headland from east to west. Under that scenario, wave surge is expected to erode boulders from the bedrock exposed at sea level, now below us at the north point. Maximum energy should be expended on boulders nearest to the headland, shifting them in a westerly direction. Wave refraction into Ensenada Almeja can be correlated with a loss in energy southward, meaning that less energy would be available to transport only smaller and smaller stones. When did all this activity begin to have an impact at San Basilio? Degradation of the volcanic structures in Ensenada San Basilio was sure to have begun not long after they first formed in mid-Pliocene time. Continued erosion both within Ensenada San Basilio and the headland at Ensenada Almeja is certain to continue with the arrival of the next hurricane. The question left to ponder is whether the severity of hurricanes in the Gulf of California is likely to increase during coming years.

Resuming the loop circuit, we start to hike through desert brush with scattered palo adán (*Fouquieria diguetii*) and low torote trees (*Bursera microphylla*) traversing a gradual rise in elevation to 82 ft (25 m) above sea level over a distance of 1,000 ft (~300 m). From this point, we can view the south end of Ensenada Almeja below to the west and a lobe-like extension of land to the east (map 3). Continuing onward to the south and climbing yet higher to an elevation of 140 ft (~43 m), we pass beyond a small, eastward-facing bay and come to a spot where it is possible to look across to sea cliffs on the opposite side. The view (figure 4.10) provides a classic look at a normal fault, with upthrown side to the east and downthrown side to the west, having a vertical offset of 56 ft (17 m). The uplifted side of the fault toward the end of the eastern promontory is distinctly dome shaped. Consulting the map (map 3) once again, the fault trace is interpreted as parallel to the fault along the east side of Ensenada Almeja.

Climbing gradually upward along the ridge parallel to the north-south valley on our west flank, the bedrock is dominated by the same banded rhyolite with a reddish cast encountered earlier at the side

Figure 4.10. View north over a limb of the eastern headland at Ensenada Almeja cut by a normal fault (see map 3). Photo by author.

of Ensenada Almeja (figure 4.9). On reaching an elevation of about 160 ft (~49 m), however, the rocks at our feet change dramatically from rhyolite to conglomerate composed of eroded rhyolite boulders and cobbles cemented in a matrix of limestone. These differ markedly from the Holocene boulders in a semicircular deposit on Ensenada Almeja. To start with, the Holocene deposit consists of loose boulders with rough edges and sharp angles, whereas the conglomerate contains smaller pieces of banded rhyolite no more than 20 in (~50 cm) in diameter that are worn and better rounded. The limestone-filled interstices among the cemented clasts are fossiliferous. For example, the external mold of a large bivalve (figure 4.11) is exquisitely preserved in a position indicating that the mollusk was wedged in place among cobbles. The fossil belongs to a species of clam (*Periglypta multicostata*) still alive today in the Gulf of California. It is the largest venus shell that occupies coastal settings from western Mexico to Peru, known to be restricted to a sandy substrate nestled among rocks at extreme low tide.[11] The distinction

Figure 4.11. Conglomerate with small boulders and cobbles eroded from banded rhyolite exposed on the ridge top above the valley linking Ensenada Almeja with Ensenada San Basilio. The loose piece of conglomerate includes the external mold of a large bivalve (*Periglypta multicostata*) preserved in the limestone matrix (pen for scale = 5.25 in, or 13.5 cm). Photo by author.

is significant because it lends evidence in support of a former rocky shoreline at this locality. In contrast to the shell coquina from the sedimentary ramp below the Spanish contessa's house (figure 4.5), the mollusks preserved in this conglomerate were buried alive in a setting close to sea level. Less than a half mile (0.70 km) away, the sedimentary bridge and related ramp explored previously represents a very different environment that developed in deeper water.

Close by and within sight, a small concrete reservoir for water is serviced by a road that leads south and then descends sharply to the west to join the north-south road in the narrow valley linking the two bays. At the bend in the road, an embankment exposes strata composed of interbedded siltstone and volcanic ash. Among these layers, spherules of black volcanic glass, commonly known as Apache

Tears, are abundant. It is likely this source was used by indigenous people to fashion small cutting tools struck off as obsidian flakes. Middens marking the encampment of nomadic visitors are close by on the north shore of Ensenada San Basilio.

From the bottom of the hill, the road follows the valley for a half mile (0.8 km) before intersecting with Ensenada Almeja and the starting point of the day's circuit. Along the way, we pass three small outcrops of dune sandstone, each not more than 4 square yards (3.3 m²) in area, tucked off to the side of the road on its east side. Taking a moment to stop at the last of these, a piece of sandstone is hammered free of the exposure. Using a hand lens to examine a fresh surface, it is found that the sandstone is dominated as much as 80 percent by fine grains of calcium carbonate that owe their origin to Pleistocene clams in Ensenada Almeja. Dark minerals eroded from rhyolite contribute to the rest of the sandstone, giving it a peppery look. It requires only a few steps beyond the outcrop to reach the modern sand dunes behind the beach. Here at the conclusion of our ramble, the same experiment is repeated by using the hand lens to examine a handful of dune sand. The ratio of beige to dark grains is about the same, suggesting that the seasonal north winds affecting the Gulf of California on an annual basis have a long history.

Third Day (visit to Rancho Santa Ana on the south shore of San Basilio Bay): There are two ways to reach the small ranch from camp. A winding road on steep hills over challenging terrain threads its way inland before emerging through the yard of the ranch corral and thereafter the cabin on the shore. By kayak (see map 3), the route is more direct along the shore to the southeast for a distance of 1.5 miles (2.5 km). The advantage in making the day's excursion by boat or kayak is that the geology exposed in sea cliffs is best appreciated from the water. Launching from the beach at camp, a sizable hillock some 165 ft (50 m) in height rises abruptly from the water's edge barely 220 yards (200 m) south. Viewed from offshore, the hillock is asymmetrical in profile with sandstone layers at a low angle forming the north side.

Passing below the cliffs toward wetlands where a typically dry drainage network converges on San Basilio Creek (map 3), the rocks bear some resemblance to features previously encountered during the first day. In particular, the same style of vertically fractured rhyolite observed close to camp and in the sea stacks below the Spanish contessa's house are well exposed at the far end. What brings joy to a geologist is the recognition of recurring patterns in the rocks that underlie any given landscape. A reasonable argument can now be made that at least three small volcanos rose above the waters in the northwest part of Pliocene San Basilio. Rocks exposed upstream around the corner of the sea cliffs are distinctly beige in tone, and much of the low-lying drainage area can be explained by easily eroded volcanic ash.

Paddling across the calm waters of the inner bay, progress is swift for the next third of a mile (0.5 km) aimed at passing between a spur of land projecting from the shore and a small island (map 3). The island's framework is dominated by the same sort of sandstone observed at the close end of the beach at camp. Onward and closer to shore, the geology is complicated by relationships among sandstone layers, igneous rocks that have the appearance of banded rhyolite similar to our previous day's encounter, and volcanic ash. Tucked behind a low hill, the cabin at Rancho Santa Ana finally comes into view, and we land on the broad beach east of the building. The beach is equal in girth to the beach at Ensenada Almeja, but more open due to the low-lying bluffs at opposite ends. The beach sand is notable on its own account, being massively influenced by carbonate grains derived from mollusk shells. A handful of sand scooped up from the beach and examined under a hand lens confirms that individual sand grains are the dominant product of ground-up shells, with only minor input of dark minerals from igneous rocks. In places, the upper beach is practically paved by shells of the chocolate clam (*Megapitaria squalida*). At some future time, these sands will compact and become lithified not as sandstone, but as a fine-grained limestone.

Kayaks pulled securely ashore, we cross the beach to the southwest skirting the ranch cabin, and continue a short distance inland (map 3). An open area free from desert brush rises from the side of an arroyo to reveal bedrock scrubbed clean by episodic floodwaters. Here, we encounter for only the second time the odd-looking, weathered surface of igneous rocks (figure 4.12) like those scrubbed by the tides around the "Birthday Cake" sea stack. These are nothing like the vertically fractured rhyolite we have come to associate with volcanic necks or the banded rhyolite characteristic of dome structures alongside Ensenada Almeja and its related promontories (figure 4.10). Unlike the sea stack, where the igneous rocks are overlain by limestone with abundant pecten fossils, the igneous rocks here give way to sandstone. It is a puzzlement. No immediate thoughts come to mind as to what transpired here long ago. Sometimes, a thing must be left to itself before the first notion of an explanation can be lofted like a trial balloon.

In plain sight, sea cliffs at the opposite end of the beach are less than a half mile (0.7 km) away to the east. The sun has climbed higher in the

Figure 4.12. Natural outcrop along an arroyo wash near the Rancho Santa Ana cabin. Lumpy rhyolite strongly weathered is overlain by sandstone. Photo by author.

sky, but there remains ample shade below the bluffs. The opportunity to inspect more rocks pulls the inquisitive naturalist in that direction, where a midday respite also may be enjoyed. A set of parallel faults is likely, given the lay of the land and the way in which the beach is pocketed between sea cliffs edged by arroyos that descend parallel to one another from the back country (map 3). Jutting outward to the bay, a 130-ft (40-m) high formation shows a clean wall on its west face. Coming into the shade, the sensation of cooler air is immediate, but it takes a few moments for the eyes to adjust to the spectacle of a magnificent geologic canvas. The wall is like a modernist painting, with angular shapes in varying shades of gray interlocked with one another (figure 4.13). Under close examination, ghostly traces of parallel lines

Figure 4.13. Vertical wall south of the cabin at Rancho Santa Ana exposing the intricate pattern of resedimented hyaloclastite representing a peculiar form of rhyolite from magma injected by stages beneath a preexisting cover. Photo by author.

are inscribed across many of the hat-size shapes. Those lines, however, are entirely out of kilter with the lines in adjacent pieces. The rocks are igneous, and the faint lines signify accrual of material in the thinnest of layers one at a time. It is the disjoined aspect of the whole deposit that is unsettling to the eye.

An analogy that comes to mind is that of a spring thaw and breakup of river ice that accumulated in layered sheets throughout the winter. Rising meltwater that flows beneath the ice causes it to buckle and fragment into jagged pieces that press together in a great mass on reaching an obstruction in the river's course. The scenario fits with a vision of energy expended explosively and with great force. The rocks before us were not formed under freezing temperatures. They were molten hot, but began to cool initially in a planar fashion throughout. The analogy works in the sense of a dramatic breakup that dismembered the layers and cast them into a chaotic mix.

The fact is that San Basilio is one of the very few places in all of Baja California where these peculiar rocks are known to occur (Johnson et al. 2019a). Indeed, they represent another class of rhyolite that volcanologists call "jigsaw fitted" hyaloclastite. The first part of the term is readily understood. The rock surface before us is like a giant jigsaw puzzle with an impossibly intricate pattern devilishly presented in muted tones of gray. Hyaloclastite is a technical term that refers to felsic lavas rich in magnesium and iron along with a high-silica content. On extrusion, they are quenched by groundwater or seawater to form thin sheets. Related to rhyolite, this kind of magma may be forced out beneath preexisting seafloor sediments like clay or silt that form a cap or cover. Pressure builds as more magma is injected from below into an expanding dome beneath the seafloor. That pressure, together with steam explosions, cause the accumulating layers to buckle and fracture. Coming from a geologist interested in the process by which rocks are disassembled by natural processes, it can be said that the hyaloclastite is resedimented. Essentially, the jigsaw pattern before us represents a physical transition from an igneous state to a secondary breccia of large but rough-edged clasts.

Sitting in the shade of the cliff while eating lunch, the realization dawns that hyaloclastite deposits also correspond to places where the rhyolite has a lumpy appearance (figure 4.12). Peering around the

corner of the outcrop along the shore to the east, the thick hyaloclastite section at this locality is seen to dip beneath towering strata
of tan siltstone and sandstone. Another wide beach appears beyond,
but comparable sea cliffs formed by siltstone and sandstone line the
rest of the bay to Punta Mercenarios a full mile (1.6 km) away. Separated by beaches, a thick but discontinuous overburden of siltstone
and sandstone encircles much of the bay. Why do such strata stop
short at the bay's edge? Except for small volcanos recognized solely
on the basis of their eroded volcanic necks that rise above the surface,
San Basilio Bay is an empty expanse that occupies a crescent-shaped
basin covering 1.5 square miles (4 km^2). How is it that any coastal
embayment occupies a particular stretch of shore anywhere around
the Gulf of California? The same question may be posed regarding
any shoreline anywhere in the world.

Reclaiming the kayaks, a slow return to camp is in store for the
afternoon. Hugging the shore, every nook and cranny is an option
for exploration. Close beyond the Rancho Santa Ana cabin, more
of the oddly eroded rocks now known to be hyaloclastite are easily
recognized in the sea cliffs. The extent of sedimentary rocks like siltstone and sandstone also demand greater attention. Lastly, bluff tops
draped by white volcanic ash make a stronger impression than during
the outbound trip. The landscape's why and wherefore take on a
greater part of our consciousness as the day's ramble is completed.
As alluring as the "magic water" of San Basilio is on its own account,
the stage on which it performs its daily spell has its foundation firmly
built on the surrounding landscape.

THREE RHYOLITES INTERWOVEN TO COMPLETE THE STORY

The task of the all-round naturalist embodied by geologist, paleon-
tologist, and ecologist is to piece together a coherent story that pulls
us into the past based on a thorough understanding of the present
world. Ensenadas San Basilio and Almeja are surrounded by a desert
landscape half dominated by sedimentary rocks that include siltstone, sandstone, conglomerate, and limestone. Fully the other half
of the countryside is filled by variations on igneous rocks that fall
under the name rhyolite. The district's three rhyolites include domes

of banded rhyolite, spires of vertically fractured rhyolite, and (unique to the entire peninsula) deposits of hyaloclastite looking akin to the pileup of river ice during the spring thaw. A substantial component of volcanic ash also lends its character to the panorama. All these rocks may be read as separate events that add up to a single epic story.

Accumulation of Pliocene silt and sand on the seafloor at San Basilio is taken as the story's starting point. Mud from the base of siltstone cliffs east of Rancho Santa Ana studied for its microfossils revealed an assemblage of 26 species belonging to bottom-dwelling, single-celled organisms with tiny shells called foraminifera (Johnson et al. 2019a, table 1). Among them are two species (*Bolivina biocostata* and *B. interjuncta*) that account for the dominant populations in the assemblage. Comparison with similar species today suggests these small marine animals dwelled in the upper bathyal zone at a water depth of nearly 500 ft (~150 m). It was beneath the muddy seafloor on which these foraminifera lived that the first injection of magma occurred, resulting in production of hyaloclastite. In part, the present shape of Ensenada San Basilio is due to the explosive nature of domes in which these shattered rocks formed below the marine surface and other domes of banded rhyolite that subsequently swelled at the surface. Certain exposures around Rancho Santa Ana show that surface flows of rhyolite are interfingered with layers of sandstone. Intervals of calm during which sedimentary layers were deposited were violently interrupted by periods of volcanic activity.

The ridge north of the contessa's house signifies a lengthy interlude of calm, during which a rocky shore appeared with boulders and cobbles eroded from banded rhyolite to make a distinctive coastal conglomerate. Dissected by crossing faults that make the correlation more difficult, the sedimentary bridge and ramp structure below the contessa's house probably developed at roughly the same time as the rocky shoreline higher on the ridge. Vertically fractured igneous rocks that solidified in the necks of small volcanos represent the third variety of rhyolite that emerged in scattered localities around the margins of the flooded San Basilio basin blown out by earlier and more effusive volcanism. Multiple examples of small volcanos from this stage in the landscape's history are evident (map 3). Even during an interlude of relative calm, accumulation of limestone around volcanic slopes was interrupted by minor injections of volcanic ash. A final great blast scattered volcanic ash across a broad swath of land west of Rancho Santa Ana.

Spires of vertically fractured rhyolite standing below the Spanish contessa's house and the southeast corner of the bay at Punta Mercenarios bear silent witness to the many storms in post-Pliocene time that stripped away the volcanic cones formerly wrapped like layered cloaks around volcanic necks. Contemporary storms are known to lash out at Ensenada San Basilio, as occurred during Hurricane Odile in 2014. The upturned sedimentary ramp below the Spanish contessa's house is a last vestige of one such cone. Its story is proclaimed to all who would look. The Holocene boulder deposit obstructing Ensenada Almeja represents a more recent span of time. It, too, has an urgent story to tell about the ongoing passage of great storms. Departing Ensenada San Basilio by boat after my first encounter in 2001, I knew that I would return and do my best to read the rocks telling the story of a truly enchanted place (plate 4).

5

Isla del Carmen and Storms over the Pliocene Tiombó Mega-delta

For the biological or the geological scientist,
it is still a land where new discoveries can be
made, where he can still be the explorer.

Joseph Wood Krutch, The Forgotten Peninsula (1961)

SIXTH LARGEST AMONG more than forty islands in the Gulf of California, Isla del Carmen covers 55 square miles (143 km^3) of rocky terrain with a peak elevation of 1,350 ft (412 m) above sea level. The island sits across the 10-mile (16-km) wide Carmen Passage from Loreto, a town of 20,000 residents and home of the founding mission church and starting point of the Camino Royal, linking all later missions in Baja and Alta California. Prior to the arrival of European priests to the settlement in 1697, the blue lagoon at the north end of the island at Bahía Salinas was visited regularly by indigenous people for the collection of salt. As a vital resource, salt harvesting continued under license to the Spanish crown during later years of colonial occupation. The operation turned into a thriving commercial business, with mechanized means of surface mining supported by a workforce living in their own village up until 1978, when operations ceased.[1] Granted the status of a Natural Protected Area in 1996, the island became part of the Loreto Bay National Marine Park, which includes a cluster of four other islands. Conservation measures safeguard terrestrial and marine ecosystems both on land and in waters covering 80 square miles (206 km²) around the five islands. Introduction of the desert bighorn sheep (*Ovis canadensis*) supports a more recent enterprise on Carmen that caters to hunters who pay a substantial fee for the privilege of bagging one of these imposing animals. Otherwise, the island has no permanent human residents, and the sheep are left to propagate outside the hunting season. Beaches at the island's more sheltered south end receive regular traffic by kayaking groups throughout all but the hottest summer months of the year.

Close by the airport with international connections and a town with a long and storied history, one might suppose that little of scientific interest remains today to discover on any of the islands in Loreto Bay. Based on his experience in the 1950s, Joseph Wood Krutch proclaimed that those devotees to the botanical and geologic sciences still had fertile ground to explore and make new discoveries almost anywhere on the peninsula. His testimony in *The Forgotten Peninsula* (Krutch 1961) appears to preclude women scientists, who long since joined the ranks of field biologists and geologists to make important contributions. My introduction to the geology of Isla del Carmen occurred in January 2003 when, with an authorization for approved scientific investigations in hand, our small group circumnavigated the island's perimeter to reconnoiter all coastal limestone exposures. Across from Loreto Airport at our final encampment during that visit, time was marked by the arrival of regularly scheduled flights. City lights glimmered brightly all night long. It was a different matter camping on the island's east coast. There, one could imagine being on the dark side of the moon in regard to contact with civilization. Given its oblong shape with a length of more than 18 miles (30 km), it requires a determined journey by boat from Loreto to reach either the north or south ends of the island. Few people venture to the central east coast of Isla del Carmen, and it remains a place rich in potential for fresh exploration and new discoveries.

The single most expansive deposit of limestone on Isla del Carmen is open to the sea from a place on the east shore called Arroyo Blanco. A month of mapping was devoted to this single locality the following year in January 2004. Our team found that a continuous succession of limestone layers is exposed in canyons leading inland for more than a half mile (~1 km). On analysis, marine fossils from these layers were shown to span much of the Pliocene and Pleistocene Epochs (Eros et al. 2006). Geologists who visited Isla Carmen earlier in 1940 recorded substantial new information regarding limestone at Marquer Bay on the opposite west coast. Prior to our visits, however, little about the geology of Arroyo Blanco was revealed and even less regarding an extraordinary "gravel deposit" adjacent to Arroyo Blanco that occupies an equally thick sequence.[2] With provisions delivered to us by boat for a week at a time, there was scarce opportunity to explore adjoining shores outside the canyon. Coming and going from our study site, we cajoled our boatman to give us an extra tour of the adjacent rocky shores whenever possible. In contrast to the limestone, the immense gravel deposit was an enigma that stayed in the back of my mind as a troubling mystery. The object of this chapter is to dispel that enigma, a story with ramifications far beyond Isla del Carmen and beyond the mountainous highlands forming the spine of peninsular Baja California. It amounts to what might be called a "shaggy-dog" story, which is the best kind of story there is to tell.

FEATURED EVENTS
Arroyo Blanco Limestone and Tiombó Conglomerate

The topography around Loreto and its neighbor town of Nopoló including Isla del Carmen covers a region of approximately 155 square miles (400 km²), which is essential to understand the system of interconnected arroyos that drain separate watersheds off the eastern slopes of Sierra de la Giganta (map 4). On the whole, the district is dominated by igneous rocks assigned to the Upper Oligocene to Middle Miocene Comondú Group, which consists mostly of andesite flows and volcanic ash layers with a minor component of fluvial and dune sandstone (Umhoefer et al. 2001). Rising abruptly nearly 3,000 ft (~915 m) above sea level, the sierra's escarpment is separated from the coast by a wide coastal plain. Viewed from the flats along Mexico Highway 1 or from the waters in the adjacent Carmen Passage, the

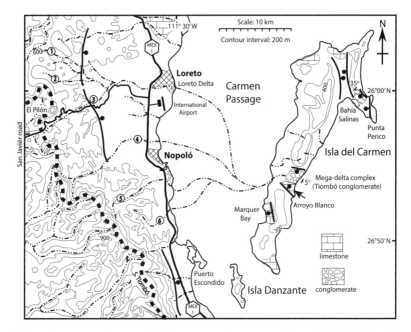

Map 4. Topography around the towns of Lorteo and Nopoló, including the Sierra de la Giganta to the west and Isla del Carmen to the east. Long and short dashed lines mark the region's primary drainage system, with numbers 1–6 denoting key watersheds. The bold dashed line traces the peninsular divide in the Sierra de la Giganta, locally identified as the San Tiombó. Overall, the district is dominated by igneous rocks assigned to the Comondú Group consisting mainly of andesite. Areas north of Loreto and on Isla del Carmen feature sedimentary rocks, including limestone and conglomerate.

sierra's colorful landscape has the look of tall ramparts with projecting battlements. Indeed, "castellated" is an adjective often applied to the Mountain of the Giant that owes its towering height to tectonic uplift along the great Loreto Fault.

The view eastward from Loreto's charming *malecón* is no less imposing for the panorama offered by Isla del Carmen. An early riser by habit, I am irresistibly drawn to the esplanade with the express purpose of witnessing sunrise over the island's dark and jagged silhouette, when a burst of light shines like a beacon to illuminate the Carmen Passage in the still morning air. The island's profile is spikey except for a line of level ground located toward the south end. It is a small detail, and most of those who enjoy the *malecón* during the early hours of the day have never paused to take notice. However, I have stood in the middle of the island on that plain, where it is possible to gaze back at the Sierra de la Giganta behind Loreto in one direction and out over the open Gulf of California in the other. I know how the level surface in the island's otherwise uneven profile got there. Two days are required to do justice to the topic at hand. The first requires the hire of a boat to get to Arroyo Blanco on the far side of Isla del Carmen from Loreto, and the second requires the hire of a car to drive the road up into the Sierra de la Giganta leading to the mission church at San Javiér (see map 4).

First Day (visit to Arroyo Blanco on Isla del Carmen and its wave-cut terrace): Departing from Loreto harbor at dawn when the sea is typically calm, the distance across the Carmen Passage on a southeast heading to intersect the island at a point slightly more than midway along its length amounts to an easy run less than an hour over a distance of 9.75 miles (~16 km). The heading is intentional, so as to approach close to the west shore at a place directly opposite Arroyo Blanco from the east shore. The coastal rocks along this part of the island rise precipitously from the water and exhibit the red color evinced by the dominate andesite belonging to the Comondú Group as found in the Sierra de la Giganta, but lacking the pale tones of volcanic ash. The boatman alters course for a run of 7.5 miles (~12 km) along the shore to the southeast end of the island. On reaching Marquer Bay (map 4), the red rocks give way to massive limestone cliffs rising more than 50 ft (~17 m) in height that represent a Pliocene embayment formerly etched into the island's flank with a sandy bottom consisting of eroded coralline red algae.[3] In the distance beyond, the same andesite rocks reappear to mark the southern boundary of the former bay. Rounding the southwest corner of the island marked by a navigational light tower, the Miocene igneous rocks are readily observed to have been cut by a set of at least three marine terraces. There is a parallel story here regarding the initial sinking of the island during Pliocene time to accommodate a greater Marquer Bay followed by a Pleistocene uplift of the island during which the terraces were carved like great steps in the older andesite rocks.

But the day's appointment is elsewhere, and we hasten to continue another 2 miles (3.75 km) around the sheltered south end of Isla del Carmen, where Pleistocene limestone formed by the same kind of coralline red algae at Marquer Bay are well exposed in low cliffs. We bypass a beautiful stretch of sandy beach composed of the same carbonate materials derived from modern coralline red algae[4] to round the southeast corner of the island where a colony of sea lions (*Zalophus californianus*) has taken up residence among the andesite rocks at the far end of the beach. The sound of barking males is muffled against a northerly breeze that has sprung up parallel to the east shore. Our boatman changes course, once again, for the final 6 miles (~9.5 km) on a northwest heading that brings us past more Comondú cliffs on the final approach to Arroyo Blanco. Our arrival is dramatic for the abrupt opening through blindingly white limestone cliffs that flank a deep channel ending at a steep beach (map 5). Surface waters stirred by a fresh breeze out of the north are suddenly quieted in the shelter of the 200-ft (~65-m) wide passage leading inland for nearly 600 ft (~180 m).

It is a small detail, but my eyes are drawn to the wall on the north side of the channel as the boatman cuts the engine and our craft glides noiselessly toward the beach. Here, the lower flange of limestone that projects out over the water is embossed by evenly spaced runnels where rainwater has etched a vertically oriented pattern into the rocks (figure 5.1). The feature is characteristic of a karst setting in which limestone is subject to slow dissolution. Although we are in a desert environment where fresh water would seem to be in short supply, the rills bespeak otherwise of intervals with a surfeit of running water. The channel itself proclaims its origin as the seaward end of a deep canyon cut downward into the limestone by floodwaters. The contradiction of excess running water in an otherwise arid place clouds my mind as the boat's prow bites gently into the sand under a rising tide. It is midmorning and we will explore the coastal cliffs north of the arroyo canyon until midafternoon, well before low tide.

Once safely disembarked on the beach, it is possible to inspect the opposite canyon walls for any signs of an offset in layering that might indicate a fault. Approximately 30 ft (~9 m) of massive limestone with identical characteristics make up the canyon's facing walls,

Figure 5.1. Lower north wall of the channel along Arroyo Blanco on the east coast of Isla del Carmen. The cliffs are 30 ft (~9 m) high and the beveled surface in the lower 6.5 ft (2 m) is distinctly eroded with downward-oriented runnels. Photo by author.

which eliminates the possibility that the drainage follows a fault line. Among the lowest-exposed parts of the Arroyo Blanco limestone (map 5, stop 1), this major cliff-forming unit is dominated by the crushed debris of algal rhodoliths similar to the Pliocene limestone on the west side at Marquer Bay, but bleached entirely white. On close examination using a hand lens for magnification, the small white fragments of broken rhodoliths account for more than half the volume of the rock, whereas the rest is divided between a lime matrix and fine black sand derived from the breakdown of andesite. The combined effect has a salt-and-pepper look wherein the white fragments are coarse compared to the black sand. Whole rhodoliths with a diameter of 1.5 in (~4 cm) are rare as body fossils within the limestone, in addition to well-rounded but scattered pebbles of andesite. Fossil marine invertebrates are mostly absent from this unit, but underlying beds exposed where the beach drops off include a

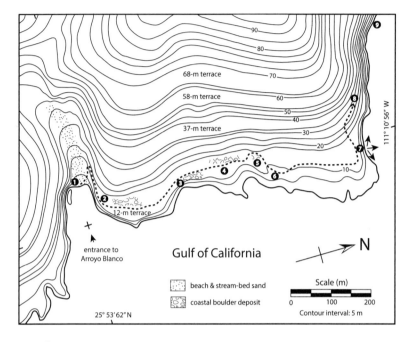

Map 5. Topography of the limestone coast that extends north from the beach at Arroyo Blanco for a half mile (0.8 km), showing a series of elevated marine terraces cut into the bedrock and various locations of geologic interest.

key index fossil represented by a distinctive sea urchin (*Clypeaster bowersi*) considered diagnostic for the Lower Pliocene.

Arroyo Blanco is the gateway to a continuous exposure of limestone layers with a composite thickness of 203 ft (62 m). Crossing from the Pliocene into the Lower Pleistocene, the full succession may be viewed only by hiking far into the canyon to its highest elevations. Access to the interior is allowed only by permit for scientific research, and our visit is restricted to the beach area and the outer edge of the first terrace along the cliff top to the north (map 5). It is worth pointing out, however, that the study by Eros et al. (2006) yielded a layer-by-layer analysis of the stratigraphic succession in which individual range zones for 22 fossil species were established with high precision. The procedure required tagging the first and last occurrence of a given fossil species within the sequence as related to the measured section from bottom to top. The local range zones for 14 of the original 22 species so studied (figure 5.2) entail fossils from the phylum Echinodermata (sea urchins and sand dollars) and from the class Bivalvia (phylum Mollusca). Patterns of overlap among these zones are sufficient to divide the regional Pliocene standard into characteristic lower, middle, and upper parts. Some sea urchins (*Clypeaster revellei*) and bivalves (*Argopecten abietis*) exhibit exceptionally long ranges, whereas others persisted for a much shorter time as a species.

The original study (Eros et al. 2006) included many more species from the Lower Pleistocene than shown here (figure 5.2). Pleistocene deposits at the top of the sequence are fully lithified as limestone, which was deposited in very shallow water. Likewise, the older Pliocene layers of limestone exhibit fossils typical of relatively shallow-water origins. Taken at face value, the overall sequence shows that the seabed went through a prolonged interval of subsidence, which agrees with the general pattern from Marquer Bay on the

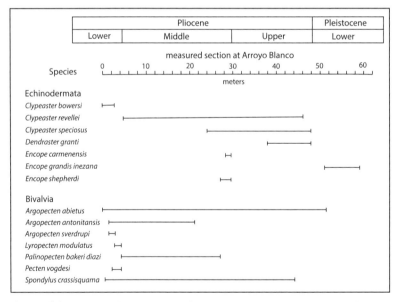

Figure 5.2. Pattern of overlapping of range zones belonging to selected species of fossil echinoderms and bivalves pegged to the measured section through Pliocene and Pleistocene limestone strata at Arroyo Blanco (modified from figure 6 in Eros et al. 2006).

opposite side of the island. The sinking of the seafloor in both areas as additional sediments accrued through time was accommodated through slippage along well-defined faults that border the limestone beds (map 4). Subsidence persisted through the early part of the Pleistocene.

Regional dynamics changed abruptly about 400,000 years ago, when Isla del Carmen began to experience tectonic uplift that resulted in the exposure of Pliocene limestone along the coast at Arroyo Blanco. Uplift was episodic, punctuated by intervals of stasis when changes in elevation slowed or ceased altogether. A series of at least 4 marine terraces was cut into the older limestone by wave action leading to development of distinct steps with outer margins left high and dry at roughly 40 ft, 120 ft, 190 ft, and 223 ft (12 m, 37 m, 58 m, and 68 m) above present-day sea level. Although eroded in preexisting Pliocene limestone, the action took place during later Pleistocene time up until about 125,000 years ago, during which certain marine invertebrates characteristic of the Pleistocene occupied the shelves. The day's hike covers a distance of 1.25 miles (2 km) round trip, ending back at the beach where our boats await (map 5). The excursion permits exploration mainly of the 40-ft (12-m) terrace, but promises to give insight to events that occurred during the early Pliocene some 4 million years ago, as well as events in the Pleistocene and Holocene between 125,000 and as late as 10,000 years ago. Spectacular views of the coast beyond await our arrival at the north end of the lowest terrace.

First, we must climb through the cliff face on the north side of the arroyo, where a faint trail allows for an assault moving diagonally over the steeply inclined surface (figure 5.3). It is a happy coincidence of lithology that the Pliocene limestone at this interval is dominated by the by-product of crushed rhodoliths. Scattered bits of bleached algal fragments the size of coarse sand protrude from the incline, giving firm traction to those with shoes having any kind of tread with a textured sole. Lending a helping hand to those behind, the 30 ft (~9 m) transit is achieved in short order.

At the top of the cliff, the trail makes an abrupt turn eastward and climbs a gentle grade parallel to the cliff edge to achieve the full height of the 40-ft (12-m) terrace (map 5). Easy walking on mostly

Figure 5.3. Rough trail ascending the limestone cliff face from the beach to the top of the 40-ft (12-m) terrace on the north side of the canyon. Photo courtesy of B. Gudveig Baarli.

level ground is in store for all but the final stretch of the outbound trek. Pausing to catch one's breath after the initial climb, the view from the rim of the canyon across the beach below to the opposite wall lends a sense of accomplishment. The canyon's south wall is seen to rise higher than our present position. Thin dark lines that extend horizontally across the south canyon wall at wide intervals consist of igneous cobbles that veiled the former seafloor during erosional events.

Flat-lying Pliocene limestone farther along the trail (map 5, locality 2) reveal bedding planes with the crowded assemblage of disarticulated shells belonging to the fossil *Argopecten* sp. (figure 5.4). A change from pure rhodolith limestone encountered in the cliff face below is abrupt, but traces of the same fragmented algal bits remain as part of the matrix encasing the shells. Individual valves are densely stacked together as a coquina that formed postmortem with the inner, convex sides of many facing upward in an unstable position. The shells were swept into place as a chaotic deposit during turbulence on the seafloor from stormy waters. Normally, upturned shells would be flipped over into a convex position by ongoing agitation, but only a few are preserved in that orientation. This shell layer was rapidly buried by a subsequent layer of rhodolith-rich sand in the lower part of the "middle" Pliocene (figure 5.2).

Farther along the trail but back from the outer edge of the terrace, the first sign of rough limestone blocks randomly piled atop the bedded surface makes an appearance. The scene almost has the look of a quarry where workmen applied crowbars to rip into the surface layers, but then abandoned their handiwork. At first, the prospect of possible human activity is only a distraction from the hunt for clean

Figure 5.4. Surface bedding plane (map 5, locality 2) in Pliocene limestone at the top of the 40-ft (12-m) terrace with abundant fossil pectens (*Argopecten* sp.). Scale 2 in (4 cm). Photo by author.

bedding planes with interesting fossils that tell more about the original Pliocene environment when those layers formed. However, the intensity of the disruption becomes increasingly clear another 275 ft (84 m) along the path, where truly enormous slabs of limestone sit in an upturned orientation torn from the outer lip of the terrace (map 5, locality 3). The incoherent jumble of limestone blocks derives from the same level of layers with abundant pecten fossils observed earlier, but also includes some scattered igneous cobbles. The lesser of these crudely rectangular blocks measures roughly 5 ft (~1.5 m) long by 4 ft (~1.25 m) wide and 45 in (~0.5 m) thick.

In contrast, the largest upturned slab at this locality was found to measure 15 ft (~4.5 m) in length by 5.25 ft (~1.6 m) wide with an average thickness of 2.5 ft (~0.75 m). Based on a small sample of limestone collected in order to make a laboratory calculation of the rock's density, it was possible to estimate the weight of this enormous block as more than 5.75 metric tons (Johnson et al. 2018). The size and weight of the largest slabs at the edge of the terrace far exceed anything that humans might accomplish with only crude tools at their disposal in a quarrying operation. Moreover, why would a quarrying operation take place here, rather than at the beach where the limestone could more easily be loaded onto a barge? If not from human activity, how then were these limestone blocks detached from the bedrock roughly 33 ft (10 m) above the current level of the sea and left in a disorderly pile?

Nature's handiwork is explained by the impact of big storms with wind-driven waves of unusual height that slammed into the limestone bluffs and drove water under intense pressure into the open cracks between limestone layers exposed at the upper edge of the terrace. It is a normal process whereby the equivalent of a powerful crowbar is applied to pry apart rock layers through hydraulic pressure. Once loosened from the bedrock, the further impact of storm waves was more than adequate to flip slabs into an upright position—or even turn them upside down. No ordinary sea storm was responsible for this work. It was a hurricane-magnitude disturbance that spiraled northward into the Gulf of California with circulating bands of wind and rain that ripped like a buzz saw from the west to east against the outer rocky coast of Isla del Carmen. Because the terrace already existed before it could be partially deconstructed by such an event, the timing of that storm (or storms) post-dates the formation of the terrace less than 125,000 years ago. Indeed, the most likely scenario for such storms would have been during the Holocene 10,000 years ago or later.

Our path follows onward another 330 ft (~100 m), crossing a remarkably clean but undulating surface of exposed limestone that reaches inland from the outer lip of the terrace for more than 80 ft (~25 m). The flats are entirely free of loose rocks, as if the surface had been swept with great efficiency (map 5, locality 4). What sits behind this open zone is an extraordinary deposit of large limestone boulders stacked 6.5 ft (2 m) high, and extending in an elongated pile for 50 yards (46 m) parallel to the shore (figure 5.5). During an earlier visit to this spot (Johnson et al. 2018), measurements in three dimensions were recorded from a representative sample of 25 boulders within the deposit. The average boulder from the sample was estimated to have a weight of 680 lbs (308 kg) and would have required the impact of a wave with a height of 14 ft (4.3 m) to make it budge. Individual boulders are, in fact, somewhat rounded in the sense that their rough edges are smoothed. Such an outcome is possible only when the blocks collide, are jostled about, and have their sharp corners trimmed. The scenario is consistent with a coastal boulder deposit (CBD), and its position and orientation behind the outer margin of the terrace speaks volumes regarding the power of storm waves to shape such a construction.

Comparison with the effect of Hurricane Odile (see chapter 4) is apt in regard to its counter-clockwise rotation as it migrated north through the Gulf of California. We can be sure that no one stood here at this place on September 14, 2014, when Odile passed over Isla del Carmen. But it would be fascinating to know if modern storms are capable of moving the big rocks when directly impacted by waves from the east. The only way to find out whether or not the next hurricane to pass this way is able to generate waves large enough to move these boulders is to tag them and check to see if they are still in the same place afterward.

The deposit at this locality is much diminished to the north, but picks up again a short distance farther on, where limestone blocks

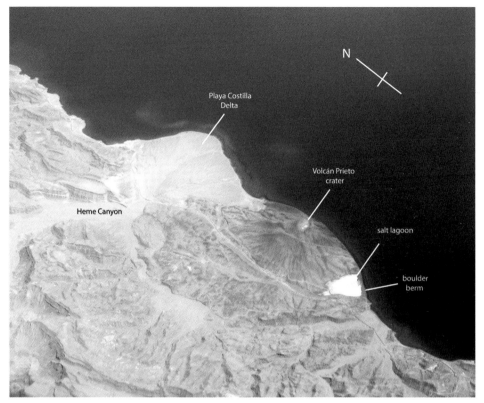

Plate 1. Aerial photo of Volcán Prieto from an altitude of 30,000 ft (9,144 m) during a commercial flight in 2016. Photo by author.

Plate 2. Field partner Jorge Ledesma-Vazquez stands in a stiff wind holding pumice cobbles on a trail of orange cobbles stranded atop the long boulder berm separating the salt lagoon at Volcán Prieto from the Gulf of California. Photo by author.

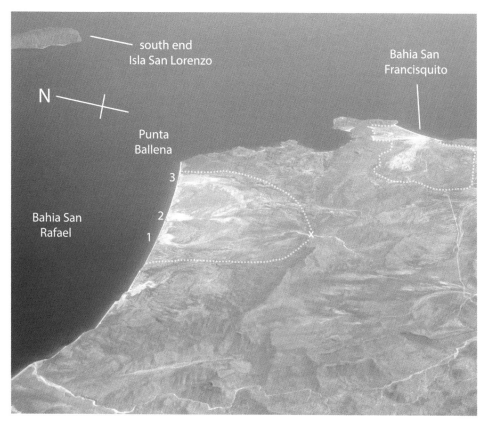

Plate 3. Aerial photo of the adjacent Punta Ballena and Bahia San Francisquito areas from an altitude of 30,000 ft (9,144 m) during a commercial flight in 2016. Dotted lines outline spaces occupied by Pliocene strata. Numbers 1–3 denote contemporary mudflats where original Pliocene strata are removed. Photo by author.

Plate 4. View over Ensenada San Basilio to the northeast with the Spanish contessa's house in the center background. Photo by author.

Plate 5. Junction of the Pliocene Arroyo Blanco limestone and Tiombó conglomerate in the delta complex on the east coast of Isla del Carmen. Uplifted marine terraces carved in the limestone are highlighted, as is the arch-shaped cross section through the delta front. Photo by author.

Plate 6. Much of Isla del Carmen viewed from an altitude of about 23,000 ft (7,000 m). Photo by author.

Plate 7. Aerial view of the inner harbor at Puerto Escondido from an altitude of about 6,000 ft (~1,830 m) with sail boats at anchor for scale. Photo by author.

Plate 8. Aerial view of Tabor Canyon looking west from an altitude below the crest of the Sierra de la Giganta. Photo by author.

Plate 9. North tip of Isla Danzante separated from the rest of the island but linked by a cobble beach visible only at low tide. Photo by author.

Plate 10. View of "The Window" from water level at low tide showing the offset of layers on opposite sides of a vertical fault. Photo by author.

Plate 11. Steep east coast of Isla Santa Cruz, showing granite extensively shot through by thick quartz veins formed by remobilized silica. Photo by Hank Ellwood.

Plate 12. South end of Isla San Diego, showing the dome-like construction of the granite island. Photo by Hank Ellwood.

Plate 13. Sculptured granite exhumed from beneath Pleistocene sandstone along the south shore of the bay at Cabo Pulmo. Photo by author.

Plate 14. Granite along the south shore of the bay at Cabo Pulmo with a resident population of an intertidal marine gastropod (*Nerita scabricosta*) characteristic of the upper tidal zone (coin is 1 in, or 2.5 cm, in diameter). Photo by author.

Plate 15. View west toward the Pacific Ocean from the southern pass across the Sierra de Laguna at an elevation of 2,592 ft (790 m). Photo by author.

Plate 16. View upstream through a gorge cut in granite on the lower slopes of the Sierra de Laguna. Arrows (white) delineate the high-water mark during flooding related to Hurricane Newton in September 2016. Photo by author.

Figure 5.5. Worn boulders piled into an oblong deposit more than 80 ft (~25 m) back from the outer edge of the lowest marine terrace at Arroyo Blanco. Photo by author.

are smaller in size and more fractured (map 5, locality 5). Observed only on one side, some of the larger blocks feature neat holes that appear as if they were mechanically drilled into the stone. Checking the intact limestone surface behind the deposit, the same holes are oriented vertically straight down into the bedrock. On average, they vary between three-eighth to three-quarter in (1 or 2 cm) in diameter and penetrate as much as 2 in (5 cm) in depth. In rare examples, the remains of thin bivalve shells are left intact within the holes. These are the "boring" bivalves, mollusks that secrete a mild acid in order to dissolve limestone (or often sandstone with a lime cement) in order to create a dwelling place out of sight within the rock.

A pholad bivalve is the technical term applied to many of these unusual marine animals, in reference to the bivalve family Pholadidae. The exact species represented by the fossil clam is uncertain, but a living species from the family Mytilidae (*Lithophaga aristata*) is known by marine biologists to bore into rocks or even the thicker shells of other mollusks in the Gulf of California.[5] In any case, the borings constitute what is called a trace fossil, and the invertebrate animals that bored into hard substrates are sometimes called "rock eaters." Of course, the pholads did not actually consume rock but behaved like most other clams and employed feeding siphons to bring food-laden water (and oxygen) from the outside. It is certain that the fossil pholads from the 40-ft (12-m) shelf at Arroyo Blanco are Pleistocene in age. They arrived when the terrace was a wave-cut platform near mean sea level that later became uplifted to its present height. Hence, the storms that ripped into the outer margin of the shelf had to have occurred sometime after these clams took up residency.

Exploring farther along the terrace to the north (map 5, locality 6), we drop below the 33-ft (10-m) level through a draw to reach a surface well worn by the waves. We are still high above the lapping sea typical on a calm day, but the rock surface is indented by shallow basins. Some are the size of a bathtub. The lower basins closer to the sea are filled with standing water, but those higher in elevation are dry. The pan-like floors of the upper basins are covered by large, cube-shaped crystals of sea salt. Salt grains begin to precipitate from seawater when most of the standing body of water has vanished through evaporation, and this is a phenomenon commonly encountered from other rocky-shore locations along the Gulf of California.[6] This spot informs us that rough seas will splash water onto the rocks above the normal tide line, where it collects in basins and remains as standing water for some time after the sea is becalmed once again. The closer the basins are to the sea, the more likely they are to be refilled with seawater on a regular basis. Those more distant from the edge of the sea at higher elevations will be flooded only seldom during a more extreme event like a sea squall. The salt-crusted pans at our feet are located well above mean sea level by as much as 23 ft (7 m). An angry sea seldom visits this elevation, but it does come from time to time. It is no wonder, then, that a hurricane might send powerful waves crashing across the 40-ft (12-m) terrace over a distance more than 80 ft (~25 m).

Climbing back onto the marine terrace, our tour continues for another 220 yards (~200 m), avoiding increasingly dense brush in order to turn a corner and reach a spot with a commanding view to the north (map 5, locality 7). The indentation in the coast where we stand is due to the juncture between the Arroyo Blanco limestone and a massive deposit called the Tiombó conglomerate (Johnson et al. 2016). The difference could not be more striking. The limestone is beige in tone, fine grained, and richly fossiliferous. The conglomerate is maroon in color with few lenses of pecten fossils found only near the base, and overwhelmingly derived from cobbles and boulders composed of igneous rock eroded from the Miocene Comondú Group. Our trek has brought us over a distance of merely a half mile (0.8 km) north from the inlet to Arroyo Blanco. The same Pliocene limestone is exposed in sea cliffs south of the inlet for roughly the same distance. From our outlook, what we see laid before us is the cross section through a gentle arch of massively bedded conglomerate separated by thin layers of sandstone that extend for 1.25 miles (2 km) to the north. In the distance, the base of the conglomerate rides up on andesite basement rocks typical of the Comondú Group. The same unconformity remains out of sight below the waterline across most of the structure in front of us. Consulting our regional topographic map (map 4), it can be appreciated that the entire package of Arroyo Blanco limestone and adjacent Tiombó conglomerate is fixed between two opposing faults that strike inland across the axis of Isla del Carmen.

The paired faults suggest that the same process of tectonic subsidence characteristic of the thick limestone succession differentiated by well-defined fossil zones (figure 5.2), also applies to the adjoining bands of massive Tiombó conglomerate. Red mountains that rise upward of 1,300 ft (~400 m) against the far horizon beyond the conglomerate show exactly where the north fault passes. But it is the arch-shaped configuration of the conglomerate layers that gives pause for thought. An interpretation worthy of consideration is the idea that the Tiombó is part of a geologic anticline. If so, the alternating layers of thick conglomerate and sandstone seams would first accumulate layer by layer like so many tiers on a multilevel cake, after which those layers would be deformed into an arch by compression from opposite sides. It is an obvious first assessment for a trained geologist, but other factors must be taken into account.

We require a better view of the Tiombó succession, the nearest cliffs of which are recessed to the west beyond the corner of the 40-ft (12-m) terrace where we stand. The ready option is to climb through the slope above nearly to the level of the next terrace in the Arroyo Blanco limestone, where we may look more closely at the adjoining conglomerate (map 5, locality 8). Although the ascent is steep and it is necessary to contend with rocks that form loose talus in places, the outlook from an elevation of 115 ft (35 m) is superior. As seen from a corner of the receding cliff line (figure 5.6), no less than eight distinct bands of conglomerate are visible. A thick apron of talus at the base of the cliffs hides the lowest part of the succession. Each of the recurrent conglomerate layers within sight is more than 13 ft (~4 m) thick. Topping it all is a layer of limestone that represents a lateral continuation of the Arroyo Blanco limestone across much of the Tiombó conglomerate. The relationships observed from this vantage help settle key questions. First, there is no fault separating the vastly different rocks of the Arroyo Blanco limestone from the Tiombó conglomerate, which must be regarded as coeval in age. Second, the source of eroded boulders and cobbles that fed the conglomerate was clearly extinguished during the final stage of subsidence, when mostly covered by limestone and sandstone.

The answer to one question often leads to yet more questions. Although the origin of limestone is apparent from the abundant coralline red algae and fossil pectens that make up those rocks, more can be learned about the conglomerate. It would help to get within an arm's reach of the conglomerate in order to appraise its texture, but the high cliff face is inaccessible. The sun has already passed its zenith as the day has advanced, and it is time to retreat to the beach where the boats await. Passing our several stopping points on the return, final impressions reinforce our overall intuition on the impact of Pliocene, Pleistocene, and Holocene events at one place on a large island.

Before setting course on the return trip to Loreto, the boatmen agree to bring us offshore close below the limestone clifftop we have traversed. They will land us for a brief stop where we can examine the

Figure 5.6. Cliff face of the Pliocene Tiombó conglomerate capped by a layer of Arroyo Blanco limestone. No less than eight massive bands of conglomerate interbedded with sandstone can be identified from a distance. The cliff base is concealed by talus, but the clifftop rises approximately 197 ft (60 m) above mean sea level. Mountains in the background are 1.25 miles (2 km) away to the north. Photo by author.

basal parts of Tiombó conglomerate exposed just above the waterline (map 5, locality 9). Talus accumulated at the edge of the water allows for a safe landing place, and affords a stable path to the base of the rock face. Here, we are in direct reach of the sea cliff. What we find is the top part of a thick band of conglomerate separated from the bottom of another band by 2 ft (60 cm) of sandstone (figure 5.7). The conglomerate consists exclusively of pebbles, cobbles, and small boulders derived from the older Miocene Comondú Group, which is dominated by igneous andesite. In both bands, the conglomerate is poorly sorted, and internal bedding is absent. On the whole, the main difference between the upper part of the first band and the lower part of the second is that larger boulders exceeding 1 ft (30 cm) are somewhat more abundant above. It means that two big loads of thoroughly mixed clasts of different sizes were suddenly dumped in the same place, one after the other separated by relative quiet at the conclusion of the first load when only sand was deposited. Substantial energy was expended in what amounts to the first of several repeated cycles recorded in the full cliff face (figure 5.7). Each of the comparatively thin sandstone beds is linked to a conglomerate band directly below, and each represents a substantial drop in energy when only sand was introduced to that cycle during its waning stage of accumulation.

Contrary to the CBDs found as a jumble of limestone blocks on the 40-ft (12-m) terrace, the Tiombó conglomerate and related sandstone units represent repetitive coastal outwash deposits (CODs). A thorough search in the lowest accessible units finds no fossils, except for lenses of sandstone bearing a few pectens at the bottom of the pile. It means that the oldest accessible part of the Tiombó succession was first deposited in seawater. During previous visits, a thorough exploration was undertaken across the flats at the

Figure 5.7. Closer view of thick bands in the Pliocene Tiombó conglomerate separated by a thin layer of sandstone (hammer for scale). Photo by author.

top of the conglomerate succession. Based on exposures in a gulley ending at the cliff face, it was found that the upper sandstone beds dip 5° to the east (map 4). Moreover, the conglomerate was found to thin appreciably halfway across the flats to the west, overlooking the Carmen Passage. With these additional facts in mind, quite another label may be applied to the CODs, part-and-parcel of which formed a gigantic delta system 1.25 miles (2 km) across from one side to the other and no less than 200 ft (61 m) in thickness on its seaward side.

Attenuated to the west, the wedge shape of the deposit is typical of a delta system that issues from a single riverine outlet, just as found with the Playa Costilla delta (plate 1). The younger delta in the north also lends credence to delta anatomy regarding development of a low arch, where most of the sediment carried to the sea descends through a central channel that builds higher and higher with time. Lesser amounts of sediment may pass through side distributaries, where the delta fan accumulates more slowly. The enormity of the Pliocene delta and its neighboring Arroyo Blanco limestone with marine terraces eroded and uplifted in later Pleistocene time is best realized by an all-encompassing view observed from a good distance offshore. Our boatmen are compliant, and the resulting view (plate 5) provides a sweeping testimony on the tumultuous events that transpired here over the last few million years.

There is much to ponder on the journey back to Loreto. Why, you might ask, are there no Holocene terraces carved into the Tiombó conglomerate equivalent to those eroded in the Arroyo Blanco limestone during late Pleistocene time? Moreover, the name for the Arroyo Blanco limestone is self-evident, but where does the name for the Tiombó conglomerate come from? The first question is easy to answer. Limestone is a strong sedimentary rock that yields to erosion but is more competent than the weakly cemented Tiombó conglomerate. That is why the vertical cliff face of the Tiombó conglomerate is recessed inland from the outer cliff face of the 40-ft (12-m) terrace in the Arroyo Blanco limestone. Equivalent marine terraces were sure to have been eroded in the conglomerate during the island's uplift, but have long since been destroyed by collapse of the poorly cemented conglomerate bands.

As for the name assigned the Tiombó conglomerate (Johnson et al. 2016), it is vital to step back for a broader geographic overview to understand where the bulk of the eroded andesite material was derived. Other conglomerate beds occur on Isla del Carmen, particularly at Punta Perico to the north (map 4). But the Tiombó conglomerate is the most massively concentrated deposit of its kind anywhere on the island. The stream valley through which its eroded pebbles, cobbles, and boulders issued is defined by the wedge-shaped deposit restricted to an east-west artery that cuts across a north-south oriented island. A limited amount of material was certain to have been eroded off the valley floor cut into the Miocene Comondú, now concealed. No contributions could have arrived from the highlands to the south that border the Arroyo Blanco limestone on the opposite side of a major fault. It is illogical to suppose that the vast part of the deposit was washed into place from the north, because in order to reach the sea at the delta front, those materials would need to make a 90° turn from south to the east. Essentially, the size of Isla del Carmen is too small to account for the high volume of the accumulated deltaic materials. While a fraction of eroded material was sure to have derived from the north valley wall, the greater part must have been transported through the valley from the west.

At first consideration, the solution may seem outlandish, as our boat reenters the Carmen Passage on a northern heading for Loreto harbor. Aside from Marquer Bay in the south and other areas at the north end of Isla del Carmen (map 4), the west shore preserves no trace of additional limestone formations. Likewise, no evidence exists for limestone anywhere between Loreto and its sister town of Nopoló on the opposite coast. The whopper of a proposition is that Isla del Carmen was attached to the mainland peninsula until the end of the Pliocene Epoch about 2 million years ago. In that case, the seabed over which we make our crossing was formerly dry land, and water courses that originated in the Sierra de la Giganta descended eastward from several different drainage areas to merge through a single valley on what now is Isla del Carmen (map 4). The peninsular divide that snakes its way through the Sierra de la Giganta (map 4) sheds water east to the Gulf of California in one direction and west to the Pacific Ocean in the other. A segment of the mountain's ridgeline bears the name San Tiombó, and that's the origin of the name for the mega-delta complex on Isla del Carmen. Could any of the andesite river cobbles eroded from the high sierra truly have made their way downstream over a distance of 18 miles (~30 km) to reach the Pliocene delta front? There is yet more to explore and consider on the peninsular mainland behind Loreto on another day.

Second Day (visit to the Sierra de la Giganta highlands and San Javier mission): The winding road that leads up through and across the Sierra de la Giganta 25 miles (40 km) to the Misión San Francisco Javier departs west from Mexico Highway 1 not far before the turn-off to the Loreto Airport (map 4). A day trip from Loreto devoted to one of the principal tributaries ending at the present-day Loreto delta and the geologic landscape through which it passes is essential to reach a more balanced appreciation for the relationship between these mountains and the big island. A range of topics including geo-morphology, geology, tectonics, and, above all, hydrology combine to make the second day's excursion attractive, no less so for the inclusion of cultural aspects associated with the church mission.

Starting in town, coordination with a morning low tide makes the trip all the more meaningful. An aerial view of the delta (figure 1.4) with Isla del Carmen in the distant background provides a useful reference for orientation. Besides Mexico Highway 1, there is a single crossing that links the north and south sides of town separated by the arroyo. From the town center, Francisco I. Madero Street intersects the protective levee on the north bank, and Manuel Pinneda Street meets the opposite levee on the south bank. The paved streets are connected by a gravel road that descends into the streambed, and that is where we choose to park our vehicle off to the side. Upstream to the west, the arroyo appears like a vast construction site in which gravel from the 140-yd (130-m) wide arroyo floor is being collected for building projects elsewhere. Leaving the vehicle, we hike along the arroyo below the north levee, crossing exposed sand bars to reach an elevated beach beyond a grove of date palms. Beaches on both sides of the channel are extensions of the levee system, and with the tide out, we are able to cross well out onto the upper part of the delta, now a third of a mile (550 m) distant from the vehicle, before reaching the gentle splash of waves at our feet.

At low tide, the modern delta exhibits two main lobes on opposite sides of a median channel that incorporates multiple sand bars. Combined, the lobes cover 62 acres (~26 ha) in a configuration typical of a classic fan delta. From this vantage, it is amusing to look across the Carmen Passage to the flat line on the horizon that represents the elevated Pliocene delta flat (figure 5.8). The maximum width of the modern delta based on its flanking beaches barely extends for more than a third of a mile (~650 m), which is only one-third the girth exposed at the Pliocene delta front on Isla del Carmen. If our intuition is correct, then the Pliocene Tiombó delta complex is the grandfather of the present-day Loreto delta. Returning to the vehicle, we begin the inland journey driving upstream along the north levee for 1.25 miles (2 km), where the road intersects Mexico Highway 1 and a long bridge crosses the arroyo to the south bank. Looking downstream from the bridge, the arroyo's path flanked by levees amounts to more than 60 acres (~24 ha). Slowing the vehicle to a crawl, we take a good look upstream to appraise the river course that extends roughly the same distance westward before disappearing around a bend to the southwest. The arroyo upstream is more narrow and unconstrained by levees. Distinct tire tracks follow the streambed, suggesting that the river course is an unofficial byway leading across the coastal plain to the foot of the sierras some 6 miles (~10 km) away.

It would require an off-road adventure to drive upstream to where tributaries belonging to the La Higuera and Las Parras arroyos converge on the main trunk of the Loreto arroyo—one leading to the northwest and the other to the southwest (map 4). It is doable, but not without shovels and a company of strong shoulders to extract a vehicle mired in sand. I have participated in too many such ill-advised exploits, and today we choose the more reliable route, taking the paved road that will bring us deep into the mountains parallel to the steeply ascending Arroyo Las Parras. On the horizon, El Pilón de Lolita rises like a bishop's miter high above the crest of the Sierra de la Giganta at an elevation of 3,300 ft (~1,000 m). Enriched in the dark mineral hornblend, the landmark is a prominent basalt intrusion that pierces a mountain ridge otherwise dominated by reddish andesite flows. According to local lore, the name derives from one of

Figure 5.8. View looking east from the modern Loreto delta at low tide across the Carmen Passage to Isla del Carmen and the horizontal line representing the uplifted Pliocene delta. Photo by author.

three sisters who should divide inherited ranch lands among them. "What is my *pilón* (slang word in Spanish for gratuity)?" asked the sister named Lolita. When shopping at market, for example, a *pilón* might be an extra orange or two added by the vendor after the agreed purchase. Lolita's reward was the bishop's miter. The peak is a popular destination for climbers, and although the view from the summit must be rewarding in all directions, we will satisfy ourselves with a roadside view out over the coastal plain from a position below El Pilón.

Leaving Mexico Highway 1, the San Javier road starts out parallel to the main arroyo, but then diverts over a series of low hills before dropping again to cross the north-south Loreto fault (map 4, location 3 also identifies Las Parras arroyo). We turn off the paved road onto a dirt track leading north a short distance to a popular swimming hole at Rancho San Telmo. Depending on the season, the pool below the cascade may be larger or smaller, but even during the warmer months, a trickle of water descends from a well-worn chute

Figure 5.9. San Telma cascade at the fall line along the Loreto master fault, where the coastal plain abuts against the Sierra de la Giganta (person for scale). Photo by author.

at the falls (figure 5.9). We have arrived at what might be called a "fall line" demarking the intersection of the coastal plain with the foothills of the Sierra de la Giganta. It is a pleasant place on a warm day, and a series of lesser water falls may be explored through the glen above the main drop. Water channeled beyond this point flows from merely one of several branch arteries all descending to the coast at Loreto. The downstream gradient is almost nonexistent, and water that spills over from the swimming hole soon vanishes as is seeps into thick gravel. Stop and consider: Is it too much of a dislocation under a cloudless blue sky to ask what the place might resemble during a major rain storm?

Resuming our journey, the paved road rapidly climbs through a succession of turns passing below El Pilón. A pullout on the outer lane after the second big switchback provides an adequate parking space, where we may alight from the vehicle for a commanding view over the canyon in which Arroyo Las Parras is entrenched. Clusters of date palms dot the streambed in the foreground, adding a dash of green to the landscape and attesting to the available flow of water. More than anything else, however, it is the enclosing walls of the canyon that speak to the power of moving water capable over time of cutting a deep gash in the rocks. The geology is laid open as a thick stack of volcanic flows, easily two dozen from the valley floor to the top of the square-looking buttes ahead (figure 5.10). The rocks are uniformly rust red in color and easily recognizable as andesite. In the distance, a horizontal streak of pale blue marks the Carmen Passage, and beyond that the long profile of Isla del Carmen stretches out nearly visible at full length. It is crucial to point out that the enormous volume of rock once filling the now-vacant space of the canyon was exported in its entirety and sent downstream by riverine erosion to clog the ribbon of streambeds all the way to the sea.

Figure 5.10. Roadside lookout over Las Parras Canyon, with the Carmen Passage and Isla del Carmen in the far distance. Photo by author.

Not only Arroyo Las Parras but also Arroyo La Higuera to the north as well as arroyos El Tular, El Zacatan, and La Santa Maria to the south empty their combined effluent to the coast within the space of 10 miles (~17 km) along the Carmen Passage. These six arroyos (map 4, numbers 1–6) account for the amalgamated watersheds linked to the Tiombó delta complex in Pliocene time. The size of the composite system amounts to the drainage of roughly 200 square miles (525 km²), including that portion now under water in the Carmen Passage that arguably occupied some 70 square miles (180 km²). It was during the early Pliocene from 5.6 until 3.2 million years ago that the precursor to the Sierra de la Giganta began to rise at a rapid rate due to tectonic uplift astride the Loreto fault (Mark et al. 2014). This occurred well within the same time frame that marine invertebrates became fossilized in the Arroyo Blanco limestone coeval with deposition of the Tiombó delta complex. In this context, the heft of the mighty fossil delta preserved on Isla del Carmen makes a reasonable counterbalance to the extensive erosion off the high fault scarp of the Sierra de la Giganta.

Introduced in chapter 1, the same interval of geologic time is attributed to the Pliocene Warm Period, during which independent evidence suggests near-permanent El Niño conditions were in effect over the adjacent Pacific Ocean basin (Wara et al. 2005; Brierley et al. 2009). Hurricane-strength storms would have been a common occurrence during that interval, more so than today. The view out over the Loreto coastal plain is impressive, but we are not finished with the day's trek. Back on the road, we climb through one last switchback before gaining the pass at the Las Parras palm grove. There, a small chapel stands on the right shoulder of the road—a resting place for pilgrims who, during the first three days of December each year, walk the route from the San Telmo bridge below to the mission church at San Javier, still some 8 miles (13 km) away. The last incline is low in relief, and soon we have crossed the east-west pass in the high sierras. From this place onward, any precipitation that falls on the west side will flow through the Viggé Biaundó valley and ultimately reach the Pacific Ocean.

Especially during the winter and early spring times, the streambed is filled with water. The Jesuit priests who established the church mission at San Javier in 1699 understood the importance of water and eventually saw to the construction of reservoirs for the irrigation of agricultural fields designed to turn the aboriginal Cochimí inhabitants into farmers. The stream meanders back and forth over sections of the road where it is reinforced by concrete. Buttes lining the pass on both sides of the road are small and low in relief compared to the great structures eroded in the same andesite flows observed from our earlier stopping point. Here, there is no great declivity cut in the tranquil landscape that surrounds us, although our elevation remains quite high at 1,740 ft (530 m). Water flowing eastward from the mountain pass reaches the Gulf of California 8 miles (13 km) away after a short but steep descent, whereas water moves more lazily westward to the Pacific Ocean over a distance of some 60 miles (~96 km).

The difference is due entirely to the physical structure of the Baja California peninsula over much of its length. In cross section southeast to northwest, the peninsula is shaped like a giant wedge on a level baseline with a long, gradual incline on one side and short but steep gradient on the other. Essentially, the initial east-west rupture from the Mexican mainland that began some 13 million years ago, acted to jack up the east side of the peninsula flush with the Gulf of California. The action intensified with the initial rise of the Sierra de la Gigante some 5 million years ago. A subsequent change in tectonics occurred roughly 3.5 million years ago, which continues today through the ongoing northwest shift of the entire peninsula congruent with the San Andreas Fault in Alta California.[7]

Constructed between 1744 and 1758, the church at San Javier is one of the finest stone-masonry structures left intact from the era of missionary outreach throughout Baja and Alta California. The seal of the Jesuit order is embossed over the main portal to the building, dedicated to Francisco Javier, known as "The Apostle of the Indias" for his missionary work as a member of that order sent to India and Japan during the middle sixteenth century.[8] The mission grounds include gardens, where a grove of olive trees was introduced and continues to thrive from its first planting. Unhappily, the native Cochimí inhabitants for whom the mission was intended were entirely killed off due to contagious diseases including smallpox brought by the Spaniards to Baja California. With no one left to serve, the mission church became all but abandoned in 1817. The annual pilgrimage, which draws participants to the restored church from across Mexico, commemorates the death of the mission's patron saint on December 3, 1552. The beautiful stone structure is left as a standing memorial to good intentions.

Closing Argument to a Tall Tale

Be it religion or science, to seek converts to any system of thought—to convince by argument from starting principles, can be a risky business. The early Jesuit missionaries to Mexico's frontier lands in Baja California were known to master the native languages of indigenous peoples they sought to influence. Some concepts were and remain more difficult to convey in translation than others. When introduced to the idea of damnation in a place of eternal heat and fire, some among the indigenous folk who lacked for much of any clothing at all were known to express a degree of enthusiasm for such an end. Who among us would elect to shiver unclothed under a cold rain, when a reliable source of heat might be guaranteed in perpetuity?

We know from historic events that torrential rainfall can and does strike the high sierras of the Baja California peninsula. We understand that flooding can and does continue to threaten the coastal plains along the Gulf of California. It might be demonstrated that the ancient Tiombó delta complex on Isla del Carmen was all too large to have formed on an island. But, is it a tale too far-fetched to think what is now an island was formerly attached to the peninsular mainland, and that riverine deposits eroded from the high sierras once crossed what is now an open channel separating that island from its former mooring? Albeit scientific in persuasion, geologists count themselves among those prophets who claim to see into the future. What can be said with surety is that Isla del Carmen now faces an undocking event all its own, whereby the island's northeastern wing (map 4) is in the early stages of separation from the rest of the island precisely along the fault lines that define the blue lagoon where seawater invades to form the extensive salt flats in the narrow valley at Bahía Salinas. When the sea level was higher during the last interglacial epoch of the Pleistocene 125,000 years ago, the valley's opposite shores were lined by coral reefs (Kirkland et al. 1966) and it would have been possible to sail a craft through the valley from one end to the other. The ongoing tectonics in the Gulf of California are complex, and the mechanics of fault-block islands defined by simple grabens and horsts is insufficient to fully account for the parallel subsidence of channels and raised islands.

Punta Perico on Isla del Carmen has its own massive deposits of conglomerate derived from the Miocene Comondú Group,[9] but they predate the Pleistocene faults that prescribe the coral reefs and subsequent valley salt flats. The same argument may be applied to the Perico conglomerate, which posits that highlands on the east side of the valley are far too restricted in size to account for the overall volume of the erosional product. Time will tell, but Punta Perico and the long ridge to its north may form its own distinct island in the future. These thoughts never fail to cross my mind whenever I am fortunate enough to fly over the big island and take in its wild expanse (plate 6).

6

Secrets of Puerto Escondido and Nearby Tabor Canyon

> Puerto Escondido, the Hidden Harbor, a place of magic.
> If one wished to design a secret personal bay, one would
> probably build something very like this little harbor.
>
> *John Steinbeck, The Log from the Sea of Cortez (1951)*

DURING THEIR EPIC voyage to the Gulf of California, John Steinbeck and his marine-biologist friend Ed Ricketts arrived at Puerto Escondido aboard the *Western Flyer* on March 25, 1940. The two became friends and confidants during the years Steinbeck resided in Pacific Grove and Ricketts operated a biological supply business out of his home and laboratory in neighboring Monterey, California. Both were to achieve acclaim in their professional pursuits. Steinbeck's breakout masterpiece, *The Grapes of Wrath*, was released for publication by Viking Press on April 14, 1939. That same year Ricketts's groundbreaking treatise, *Between Pacific Tides*, was published by Stanford University Press. A decidedly more academic volume, the latter set a new standard for how marine life was described in a holistic way as interrelated organisms regulated by variations in the physical environments where they best thrived. Previously, the practice was to describe the life cycle of a given marine invertebrate separately from those plants and animals with which it cohabited. Ricketts's insights were significant, as the concept of *ecology* was yet to come into fashion. However, scientific books seldom result in the kind of financial success that accrues to the author of a popular novel.

Steinbeck's success provided him with a degree of financial independence that had previously eluded the struggling author. A little-known fact about Steinbeck's background as a Stanford undergraduate is that he spent a summer studying general zoology at Stanford's Hopkins Marine Station in Pacific Grove.[1] Thus there existed ample scientific grounds for shared interests between the writer and the biologist. Besides, Ricketts was not the sort of person limited exclusively by the science at which he labored. He was regarded as a kind of "renaissance" personage, with interests in topics as varied as classical music and Eastern religions. With his first substantial earnings as a novelist, Steinbeck paid out $2,500 for a 6-week charter of the 76-ft commercial purse seiner *Western Flyer*, captained by Tony Berry out of Monterey. Based on inflation, Steinbeck's 1940 investment is equivalent to about $47,000 in 2020 dollars. In 1941, Viking published the resulting narrative co-authored by Steinbeck and Ricketts under the title *Sea of Cortez: A Leisurely Journal of Travel and Research*. With the entrance of the United States into World War II after the attack on Hawaii's Pearl Harbor, the book became lost in the tumult and anxieties of war fever. Steinbeck went on to immortalize Ed Ricketts as the larger-than-life character "Doc" in his novels *Cannery Row* (1945) and *Sweet Thursday* (1954). Reissued as *The Log from the Sea of Cortez* long after the war in 1951, the book's narrative portion under Steinbeck's authorship became a cult classic among ecologists. The chapter on Puerto Escondido and the pair's adventures in the nearby Sierra de la Giganta are iconic for Steinbeck's droll humor and Ricketts's boundless enthusiasm for the marine creatures he so treasured.

The magic felt by Steinbeck at Puerto Escondido occurs for a reason. The foremost goal of this chapter is to reveal the underlying geologic and geomorphologic background that makes it such a special place. Close-by Tabor Canyon is where Steinbeck and Ricketts made their famous foray into the Sierra de la Giganta on the hunt for native mountain sheep (*borrego*). The ethos eschewed by these American visitors stands in contrast to that of hunters drawn by the commercial hunting operation on Isla del Carmen today. Moreover, the canyon's physical setting is in sharp contrast to that of the water course at Las Parras that descends from the same mountains a short distance to the north (see chapter 5).

Time Travel Back to 1940 and Earlier

First Day (visit to Puerto Escondido): From Loreto on Mexico Highway 1, the turnoff to Puerto Escondido is reached after a distance of 15 miles (24 km) and is well marked by signage (see figure 1.7, locality 5). The outer harbor at Puerto Escondido covers an area of 125 acres (0.5 km²) and is perfectly shielded from prevailing winds out of the north by the quasi-island called Infirmary Hill (Cerro La Enfermería). The mouth that opens to the hidden lagoon within (La Bocana) amounts to no more than 165 ft (~50 m) in width. It is all but invisible when arriving by boat from the south (map 6). The suddenness and utter surprise of the inner lagoon to the weary mariner seeking a safe anchorage for the night yields more than a sense of relief. It is a sight that conjures total enchantment. The inner harbor is 570 acres (2.3 km²) in size, ample enough to shelter an armada of sailing vessels and yachts. On that March day in 1940, one can imagine that the *Western Flyer* may have been the sole visitor at the entrance to the inner harbor. Steinbeck tells that their charts showed a maximum depth of 18 ft (5.5 m) within, considered too risky for the *Western Flyer* on account of the tidal range. Captain Tony Berry chose the outer harbor with its deeper anchorage for the duration of their two-and-a-half day stay. The inner harbor was explored using the vessel's unreliable motorized skiff, affectionately known as the "Sea Cow."

Harbor facilities have expanded and improved over the years, and it is common to find dozens of sailing vessels at anchor in the lagoon. Larger ships that draw visitors to the Gulf of California as ecotourists often make a call at the main wharf just inside the entrance. Our first objective is to climb into the hills on the west side of the lagoon for an overview of the entire setting. Paved side roads offer easy access to the base of the hills about a half mile (750 m) from the entrance

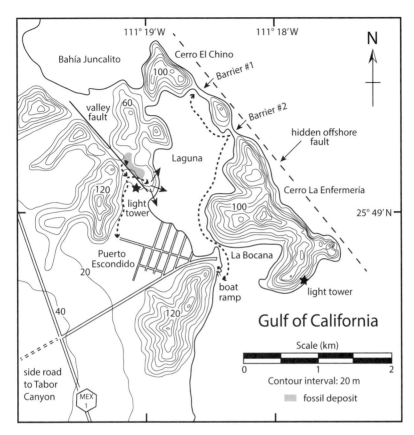

Map 6. Topography around the outer and inner (hidden) harbor at Puerto Escondido located 15 miles (24 km) south of Loreto (see also figure 1.7).

road, but thereafter it may be necessary to hike along a dirt road leading to a light tower at the side of a prominent indentation in the hills (map 6). The tower rises from a point near the distal end of a fault that cuts southward through a narrow valley from neighboring Bahía Juncalito. A saddle in the hills at an elevation of 150 ft (45 m) marks the topographic boundary between the two bodies of water. Leaving the flats behind, we ascend a trail approaching the top of the divide, where a narrower path to the east departs to a lookout on an andesite knoll at an elevation of about 155 ft (47 m) above sea level. Here, we may enjoy an unobstructed view over much of the inner harbor (figure 6.1). The elevation provides perspective that reveals

Figure 6.1. View overlooking the inner harbor at Puerto Escondido with barriers 1 and 2 labeled on opposite sides of a connecting islet. Isla del Carmen appears in the far distance across the Carmen Passage. Photo by author.

perhaps the most telling secret of the place. Isolated by a single and very narrow entrance from the south, the impressive scale of the inner harbor is seen to be well protected from winds emanating from virtually all directions. From a geomorphic point of view, the outstanding question is why such a perfect anchorage should exist at all.

Gazing to the northeast, with the profile of Isla del Carmen on the distant horizon, physical vulnerabilities in the hidden harbor are readily apparent. The outer margin of the lagoon would be open to the Carmen Passage in two places absent a pair of barriers linked to a connecting islet. The number of times I have stopped with fellow geologists and local friends escapes my memory, but on each occasion the barriers have elicited spirited discussion. Steinbeck's terse encapsulation rings in our ears: that the perfect design of the place might be replicated only by a construction under human supervision. Are the two barriers (figure 6.1) artificial, or did they accrue gradually through time as a result of natural processes? The debate advanced from time to time, aided by observations using binoculars in search of any signs like concrete retaining walls. To be sure, the length of the two barriers is impressive. The northern barrier (map 6, #1) stretches from Cerro El Chino to the central islet for a distance of 820 ft (250 m). The next barrier (map 6, #2) extends for a lesser distance of 460 ft (140 m), linking the islet with Cerro La Enfermería to the south. The size of the barriers is not trivial. Were they to be removed, the resulting gaps are sufficient to permit the passage of some of the larger craft now plying the Gulf of California. Moreover, a porous inner harbor certainly would be less sheltered from disturbances affecting the seas outside. As to their origins, any resolution to the debate requires closer inspection of the structures in question, an assignment for the afternoon's excursion by kayak.

As for the rest of the morning, additional aspects of geologic significance beg to be investigated in the surrounding area. The remains of marine shells, bleached a ghostly white, lay scattered below the west side of our promontory. What does this mean, and how extensive are the shell deposits? We retrace our steps to the juncture with the main trail, noting as we go that the pathway is strewn with shells as if placed there by human hands to highlight the passage. The cover of shells is thin, but unmistakable. For the most part, the shells represent a single species of bivalve from the phylum Mollusca (*Chione californiensis*), all of which are disarticulated as separated valves. The pathway sits well above present-day sea level. Other than a rise in sea level during an earlier time, it is difficult to envision another way that the shells may have arrived here in such profusion. From personal observation, I recall an event witnessed along the shore near Mulége when at low tide a common seagull (*Larus californicus*) swooped down on the tidal flat and plucked up a live clam in its claws. Climbing to a decent height, the bird dropped the shell onto the rocks with the result that the shell broke apart to expose the flesh within. Returning to the ground, the gull jabbed at the shell with its beak, separated the meat from the shell, and swallowed the prize. Animal behavior of this kind is astonishing, but it would take an enormous population of seagulls working hard for a very long time to produce such a spread of seashells as found on the pathway. Moreover, the underlying sand would seldom suffice to crack a shell dropped from most heights.

Rejoining the main trail (map 6), we climb to the summit of the divide to reach the head of a narrow valley below which the aquamarine waters of Bahía Juncalito may be seen in the distance. The trail is lightly used, but remains distinct and traces a straight line through the valley between the bay to the north and the lagoon at Puerto Escondido. Discovery of arrowheads in the immediate area indicates that the path has long been in use. Here on the saddle, the ground cover of vegetation is sparse, and the broad extent of shell cover is obvious. In a single sample covering a mere 250 square inches (1,615 cm²), approximately seventy-five valves of the dominant *Chione* clam may be found littering the surface (figure 6.2). The tally features whole (unbroken) valves, although fragmented shells also are present. From the sample, there is scarcely any evidence for other mollusks. The operculum of a marine snail (*Turbo fluctuosus*) is small and easy to overlook. That particular species is diagnostic for a herbivorous gastropod typically found living in an intertidal setting today. Another snail hidden in the array is a predatory gastropod (*Murex elenensis*), also found today living in the intertidal zone where it feeds on other mollusks and barnacles. The *Turbo* utilizes a hard mouthpiece called a radula like a file to scrape off marine algae critical to its diet, whereas the *Murex* uses a similar mouth device to bore a hole through the shells of its prey to gain access for feeding.[2] Rare, but easy to spot within the mix, is a small oyster (*Ostrea fischeri*), also known to encrust rocks in an intertidal environment.

Taking time to survey a wider exposure of the deposit, several additional clams may be found, all relatively rare but nearly twice the size of the dominant *Chione* bivalve. They include a turkey shell (*Cardita megastrophica*), bittersweet shell (*Glycymeris maculata*), chocolate shell (*Megapitaria squalida*), cockle shell (*Trachycardium panamense*), and rock oyster (*Spondylus calcifer*). The short list requires some effort to assemble, because these species (all of which are extant) represent a clear rarity within the overall assemblage of shells. However, all agree with the dominant *Chione* as being characteristic of a shallow-water setting. The deposit qualifies as a shell drape on account of its relatively thin cover. Indeed, using a rock hammer to dig a shallow pit in the ground, it is apparent that the deposit is rarely more than 6 in (15 cm) in thickness. Conceding that

Figure 6.2. Close view of fossil shells scattered across the topographic saddle in the foothills between Bahía Juncalito and Puerto Escondido (pocket knife for scale). Photo by author.

the veneer of shells signifies a bare wash of intertidal water, we are left with the problem of explaining how the deposit came to rest at its present elevation some 165 ft (~50 m) above present sea level inland from open gulf waters.

Like many other Pleistocene shell deposits from peninsular shores up and down the Gulf of California, the deposit above Puerto Escondido consists entirely of loose (unconsolidated) materials, and fits with the last interglacial epoch some 125,000 years ago, when sea level worldwide was approximately 20 ft (6 m) higher than today.[3] Far more commonly, however, such drapes are limited to marine terraces typically 40 ft (12 m) above sea level as widely surveyed along gulf shores between Santa Rosalia and Loreto.[4] The difference of an additional 20 ft (6 m) from the global average is attributed to tectonic uplift specific to peninsular shores since the last 125,000 years. Especially along Baja California's east coastline, the average rate of uplift is calculated to have been about 2 in (5 cm) over 1,000 years. The inescapable testimony of the Pleistocene deposit at Puerto Escondido is that it occurs at an elevation 125 ft (38 m) higher than normal elsewhere in the surrounding gulf region. Whereas the source of uplift affecting the eastern shore is related to ongoing tectonics offshore in the Gulf of California, the immediate source of extraneous uplift in the hills behind Puerto Escondido can be associated with the fault that crosses the divide between Bahía Juncalito and the inner harbor. This minor fault is trivial compared to the great Loreto fault located less than 1.25 miles (2 km) inland at the foot of the Sierra de la Giganta (see chapter 5), but was locally effective, nonetheless. The shell deposit attests to a former time before the harbor lagoon became closed off from the open gulf. In viewing the topography of the surrounding area and the trend of the valley fault (map 6), it is tempting to extend the line southeast, such that it intersects La Bocana. It cannot be denied that the narrow opening to the inner harbor is dramatic between opposing hills that rise steeply to elevations exceeding 525 ft (160 m). It is logical to suppose that the hills were once linked together before fault-related weaknesses were introduced.

The morning grows late, and we are sorely tempted to return to the overlook for an early lunch during which geomorphic issues may be contemplated in clear sight of Steinbeck's magical harbor laid out below. Prior to development of the valley fault, the ground to the west was in a much lower position barely awash with seawater. Today's sole opening to the lagoon may have been blocked 125,000 years ago by the confluence of hills with a minimum elevation of no less than 330 ft (100 m), but the two barriers (figure 6.1) did not yet exist. Indeed, the islet that connects the barriers today was surely less of an obstruction during the Late Pleistocene highstand in sea level. Biological commerce to and from the more porous embayment was abetted by a brief marine incursion that left behind the shell drape with abundant *Chione* shells. The landscape has yet other stories to be discerned. With our midday repose satisfied, it is time to return to the vehicles and proceed to the boat ramp where kayaks easily can be launched.

Payment of a small fee at the office adjacent to the boat ramp covers the cost of vehicle parking and use of launch facilities. We are about to cast off in waters that are mythic in stature, owing to the descriptions by Steinbeck in 1940. The distance to the north end of barrier #1 and return trip stopping at barrier #2 (map 6) amounts to a little more than 3 miles (~5 km). Once in the inner harbor, the water will be dead flat and the paddling effortless. But first, we must maneuver through the outer harbor, where the *Western Flyer* sat at anchor all those years ago.

It was here that an amusing encounter between wildlife and the visitors from Alta California's Monterey took place. Steinbeck and Ricketts had set about collecting biological samples from the rocky shore on the southwest side of Infirmary Hill. On that March afternoon, the air was warm and fragrant with the smell of the flowering red mangroves (*Rhizophora mangle*). One of the crew members ("Tiny" Colletto) sat in the Sea Cow offshore, when a giant manta ray (*Manta birostris*) was observed to follow a path aimed directly toward the skiff. An adult ray might weigh as much as 3,600 lbs (1,633 kg), and Steinbeck informs us that the animal in question had a wing span greater than 10 ft (3 m) from tip to tip. The ray dived beneath the skiff and surfaced on the other side. Steinbeck (1951, 158) comments that "this great fish could have flicked Tiny and boat and all into the air with one flap of its wing." The skiff's occupant is said to have sat motionless for some time contemplating the encounter, before shouting: "Did you see that Goddamned thing!"

Early in the morning on March 27, 1940, during low tide, Steinbeck and Ricketts entered the inner harbor aboard the skiff and commenced to sample the lagoon's marine life. They found that the brown cucumber (*Isostichopus fusca*), a holothurian typically 6 in (15 cm) long, dominated the marine fauna in the shallows along the eastern side. The holothurian is a member of the phylum Echinodermata characterized by an elongated body (in the general shape of a cucumber), with feeding tentacles arrayed around the mouth at one end and an anus at the other. In between all along the ventral surface are five rows of tube feet used for locomotion over the seabed. The animal is covered by soft skin, unlike its relative, the starfish, but it is the five rows of tube feet much like those of the starfish that merit its membership among the Echinodermata. A distinctive trait of the holothurian is that it expels its intestinal lining when attacked by a predator. While an easy meal waylays the predator, the sea cucumber may escape to regenerate its guts at a later time. Steinbeck and Ricketts observed many hundreds of the brown cucumber as they fed and moved slowly over the seabed. The visitors struggled to collect and preserve some in alcohol, before the captured holothurians could disgorge their intestinal tract.

Along the lagoon's west shore, Steinbeck and Ricketts encountered sand flats entirely devoid of the brown cucumber and largely sterile of other marine life as well. Closing the circuit on their sampling foray, the pair returned to La Bocana, where the sea was observed to rush inward on a returning tide. There, they found a remarkably rich marine fauna that included a profusion of green-and-red cushion stars (perhaps *Patria miniata*) and clusters of soft corals (likely a species in the genus *Palythoa*) in knobby shapes attached to the rocks. The outer mouth of the lagoon proved be one of the most diverse locations sampled during the entire expedition. It is recorded that they took sponges, tunicates, chitons, limpets, bivalves, snails, hermit crabs, as well as eight species of sea cucumbers and eleven species of starfish from this single station (Steinbeck and Ricketts 1941). The contrast between the profusion of marine invertebrates in the outer harbor and paucity of life inside the lagoon underscores the significance of the Pleistocene shells in the foothills above the hidden harbor. Although dominated by one species of bivalve, the 125,000-year-old seabed had unrestricted access to the sea before the lagoon was isolated.

Our present investigation of the inner harbor is focused entirely on the physical character of its geomorphology. Yet we cannot help but reflect on the biological survey conducted in 1940 along these very shores as we glide over still waters (map 6). One thing is abundantly clear as we pass close beside barrier #2 on the way to barrier #1. There is no trace of human construction visible from the water. The barriers are formed entirely of loose cobbles and boulders eroded from adjacent rocky shores. On reaching the north end of barrier #1, it is surprising to find that the drop-off from the shore is precipitous. The bottom of the kayak makes no scraping noise as the prow reaches the shore. One can stand on alighting from the kayak, but the increase in water depth is abrupt and one would become fully submerged only three or four steps outward from the shore.

Ashore, the barrier is found to be robust in girth. From the sheltered edge of the lagoon to the outer margin on the seaward side, the barrier width is roughly 100 ft (~30 m). The height of the barrier along its axial midline is approximately 9 ft (~2.75 m) above mean sea level. Tree limbs and other woody debris lie scattered along the top of the barrier, suggesting the flotsam arrives during gales that coincide with high tides. The most prominent geomorphic feature is the decidedly asymmetrical profile of Cerro El China as seen in cross section looking to the northwest (figure 6.3). The steep cliff face is the immediate source of all the loose material forming the barrier. Consulting, once again, the topographic map for this part of Puerto Escondido (map 6), it comes as no surprise to see that lines of elevation are more closely

Figure 6.3. North end of barrier #1 linked to the peninsular mainland at Cerro El Chino, showing a steeply eroded rocky shore and woody debris washed onto the barrier during storms likely to have coincided with high tides. Photo by Erlend Johnson.

spaced along the outer shore than the inner side next to the lagoon. The exposed cliffs make a strong impression, rising as they do at an angle of 55°. The utter absence of covering vegetation attests to cliff-side instability and the impending threat of rock falls. Elsewhere, the lesser slopes around the back side are stabilized by clumps of the coast cholla (*Cylindropuntia prolifera*) and sparse shrubs. Between patches of vegetation, talus resides at the angle of repose.

Moving to the outer shore, intermingled cobbles and boulders are tinged a soft green in tone by a film of marine algae that tells us the tide is out. Compared to the abruptness of the inner shore, where the kayaks remain stowed, the outer shore recedes ever so gently at a low angle into the open sea (figure 6.4). The boulders are formed by volcanic rocks of andesite and they tend to be oblong in shape. The largest among them have long axes exceeding 4.3 ft (1.3 m). Based on the density of andesite, the estimated weight of such a large boulder exceeds 880 lbs (400 kg). The same kind of analysis performed on the storm deposits at Ensenada Almeja (see chapter 4) and Isla del Carmen (see chapter 5) have been conducted here at the barriers of Puerto Escondido (Johnson et al. 2020). As it turns out, andesite has a greater density than rhyolite or limestone, which means that an andesite boulder the same size as a rhyolite or limestone boulder will actually weigh more. Moreover, it takes a larger wave with an expenditure of greater energy to budge such a boulder.

Based on the same mathematical equations applied in previous studies (Johnson et al. 2018; Johnson et al. 2019b), a storm wave with a height of 20 ft (~6 m) is needed to loosen and transport the largest andesite boulders from the base of the exposed sea cliffs at Cerro El Chino to the adjacent barrier # 1 (map 6). It is important to note that the barrier deposit consists of thoroughly mixed clasts that range in size from pebbles to cobbles to boulders (figure 6.4). Clearly, the energy imparted by a modest squall is sufficient to move pebbles and small cobbles, but insufficient to move boulders. Within the first 164 ft (50 m) on the seaward margin of barrier #1 starting from the southeast corner of Cerro El Chino, the largest 25 boulders surveyed were found to require an average wave height of 15 ft (4.6 m). In general, the largest boulders in the next 3 samples over the following 492 ft (150 m) were found to decline somewhat in average size and weight.

Figure 6.4. Middle portion of barrier #1 facing east with Isla del Carmen on the horizon across the Carmen Passage. The largest boulder in the foreground has a long axis of 4.45 ft (~1.3 m). Photo by Erlend Johnson.

In fact, far fewer boulders occur at the other end of barrier #1 closer to the islet with which it is linked. It means that Cerro El Chino is the primary source for the eroded clasts in barrier #1.

Evidence is circumstantial, but only storms of hurricane strength will achieve the power to carry the largest boulders from the base of the exposed cliff face at Cerro El Chino to a final resting place embedded in barrier #1. We know that hurricanes travel into the Gulf of California from time to time impacting the Loreto area, most recently Hurricane Odile in 2014 and Hurricane Lorena in 2019. Counter-clockwise rotation of wind bands during these storms can be modeled to initially strike the peninsular gulf coast from offshore with wave surge moving from east to west. As a storm passes any given location, wind direction is expected to shift increasingly to a northeast source, pushing waves in a southwest direction. Such a pattern is perfectly poised to first strip blocks from the steep cliff face and then to carry them away along the shore until they reach the barriers. Repeated storms cause the blocks to grind against one another with the result that rough edges become smoothed. To my knowledge, no one has ever attempted to occupy the Puerto Escondido barriers during a big storm. The noise of cobbles and boulders colliding with one another with the arrival of each wave would be deafening. Not in living memory have the wildest storms breached these barriers. Their primary effect appears to add ever more material to the already sturdy girth of these natural constructions.

Pattern repetition brings extra clarity to the comprehension of any phenomenon. As the afternoon has advanced, we must return to our kayaks and resume the excursion to barrier #2 by skirting the sheltered side of the connecting islet (map 6). The sheltered side of the islet rises above the water through a 20° slope. Stowing the kayaks and crossing to the middle of barrier #2, the east-west profile of the connecting islet is startling in its asymmetry. The seaward cliff face rises at an angle of 50° from the water's edge. Hidden from view from our earlier lookout above the lagoon this morning (figure 6.1), the contrast now seen in the islet's east face appears as if a giant took an ax to the structure and cleaved off a huge piece. As observed at Cerro El Chino, no stabilizing vegetation exists on the exposed, steeper side of the island. Likewise, abundant woody debris covers the barrier's midline. The powerful giant is the sea itself, often benign in mood but sometimes angry and severe. Moving to the east edge of the barrier, much the same mixture of andesite pebbles, cobbles, and boulders is found in the tidal zone. As before on barrier #1, the largest boulder within the first 164 ft (50 m) south of the nearest bedrock includes those nearly 4 ft (1.18 m) in maximum diameter. But as before, average boulder size equates with wave heights estimated close to 15 ft (4.6 m).

The boulder survey from the next 164 ft (50 m) confirms diminishment in average size and estimated weight based on the density of andesite (Johnson et al. 2020). Fewer and fewer boulders of any appreciable size are found on the barrier where it connects with Cerro La Enfermería 460 ft (140 m) to the southeast. Thus the evidence suggests that most of the andesite clasts in barrier #2 were derived from erosion of the islet to its immediate northwest. Here, we may pause to consider how much of the islet was likely to have been eroded over time and whether or not a sufficient volume was available for the construction of barrier #2. Such an exercise entails graph paper and a few back-of-the envelope calculations. The first step is to represent the area and topography of the islet as it exists today superimposed on a scaled grid (figure 6.5a). A line drawn through the map is parallel to the elliptical shape of the present-day islet and effectively divides it into two parts, with the steeper slope shown by the closest lines of elevation segregated to one side. Assuming that the islet was originally more symmetrical in cross section than today, the geomorphologist's task is to perform a reconstruction that redraws lines of topography more evenly around the islet's circumference (figure 6.5b).

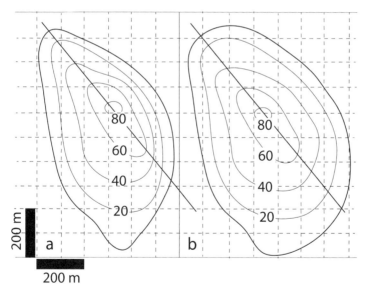

Figure 6.5. Map comparisons after (a) and before (b) scaled to estimate loss in volume over time experienced by the connecting islet between barriers #1 and #2 (see map 6 for context). Each square superimposed over the topography is 100 x 100 m on a side and represents a flat surface of 10,000 m². In three dimensions, each block in the structure is 100 x 100 x 20 m and represents a volume of 200,000 m³.

The rest requires only the elementary mathematics of addition applied to the model of a tiered wedding cake to find the volume of the existing islet in cubic units. The surface area inside each line of topography at 65.5-ft (20-m) intervals is estimated and multiplied by its thickness to arrive at a unit of volume. Repeating the process five times to arrive at the islet's summit, the serial summation of volumes comes to a little less than 500 million cubic feet (~13.85 million m³). The same operation applies to an islet restored to a more perfect symmetry with the result estimated at 680 million cubic feet (~19.25 million m³). The difference between the two sums makes a reasonable estimate of the rock volume lost by shore erosion, calculated at 190.7 million cubic feet (5.4 million m³). In easy numbers, it may be argued that the connecting islet lost more than 25 percent of its former volume due to an imbalance of coastal erosion on its exposed flank. Coastal uplift along the entire front may be imagined due to an offshore fault (now hidden from view) that led to increased exposure and further attrition. Clearly, the visible valley fault and projected offshore fault are parallel to one another (map 6).

Yet other questions remain. When and over what interval of time did much of the erosion take place? It is worthy of note that the outer, seaward shores of El Chino and La Enfermería are far more linear than their inner shores. It might well be expected that marine terraces formed during tectonic uplift, not unlike those on the east side of Isla del Carmen (see chapter 5). On a regional basis, the youngest marine terrace is the 40-ft (12-m) platform, which is dated to the late Pleistocene some 125,000 years ago. As no trace of such a feature remains outwardly at Puerto Escondido, the implied answer is that any preexisting terrace was erased by coastal erosion. As the islet between barriers #1 and #2 is demonstrated to have lost some 25 percent of its volume, and moreover because the asymmetry of El Chino and La Enfermería share much the same shape, it is logical to conclude that a significant amount of erosion occurred post-Pleistocene in Holocene time and continues through today.

Although we don't know for certain the thickness of either barrier, their width and length are easily ascertained. Using the most conservative estimate for thickness no greater than the depth of water in the lagoon, the volume of unconsolidated materials can be guessed. The contents entombed in barrier #1 amount to roughly 1.5 billion cubic feet (42.5 million m³). The shorter barrier #2 contains 812 million cubic feet (23 million m³) of pebbles, cobbles, and boulders. In effect, the barriers that ensure the tranquility of the inner harbor at Puerto Escondido might be removed and restored many times over based on the volume of solid rocks carried away by storms from its exposed outer shores during the last 10,000 years.

The afternoon recedes and we have yet to return a mile and a quarter (2 km) back through La Bocana to reach the boat ramp in the outer harbor. The day's outing has revealed many secrets held by the landscape around Puerto Escondido. Gliding through the still waters, there is deep satisfaction won from the small effort invested. To truly know a place, it must be examined from every possible angle. It is not enough to understand the making of Puerto Escondido from the vantage observed from its western hills. The lagoon's outer barriers must be encountered close-up, as well. Without exaggeration, it may be said that we've come to know Puerto Escondido both from the inside and outside.

Second Day (visit to Tabor Canyon near Puerto Escondido): Opposite the paved entrance to Puerto Escondido, there is a secondary road leading to the mouth of Tabor Canyon after a short distance across the flats (map 6). Leaving the crew of the *Western Flyer* at the outer harbor, it was here that Steinbeck and Ricketts commenced an overnight hunting expedition taking them into the Sierra de la Giganta through a narrow chasm that defies the definition of an actual canyon. Tabor Canyon is more like a ragged gouge in the side of the mountain that traces upward at a high angle, becoming ever more precipitous with altitude. A trail of sorts ascends through the mountain flank and crosses the summit at an elevation close to 3,000 ft (~914 m) above sea level. In the company of a local rancher named Leopold Pérpuly and others from Loreto, including hunters tasked with tracking native mountain sheep, the Monterey pair camped overnight with their hosts at the side of a mountain pool halfway up the escarpment.

Seen from the Carmen Passage offshore Puerto Escondido, the immensity of perfectly layered rocks exposed across the mountain front presents a formidable if not intimidating face. Cleaving northwest-southeast close along the base of the mountain, uplift across the master Loreto fault has left the imprint of colossal tectonic forces. Like Puerto Escondido itself, the canyon carries a mystical air about it, owing to Steinbeck's description of the mountain's physical grandeur and the local citizenry in whose care they found themselves. Later visitors who marveled at the mountain's stratification and wondered at the composition of those lofty layers hardly can be blamed for supposing that water-laid sedimentary rocks are involved.[5] Instead, volcanic rocks formed by repetitious layers of andesite lava and ash falls (tuff) make up the greater part of the escarpment. Some sandstone occurs near the base of the Sierra de la Giganta. However, the sandstone is poorly exposed in the canyons west of Loreto (see chapter 5) and, hence, the incision at Tabor Canyon is well suited to an exploratory trip to access the relationship between sedimentary and volcanic rocks in the mountain's makeup.

Starting in the cool of the morning under an open sky, twists and turns in the course of the arroyo on ascent are sure to provide pockets of shade from the sun's growing intensity. The hike could consume much of the day up to and back from the summit, but we are content to reach the rock pools part way up, where Steinbeck and Ricketts were likely to have camped.

At the end of the road, we are confronted by a massive diversion channel built at a right angle to the mouth of Tabor Canyon (figure 6.6). The channel itself is nearly 50 ft (~15 m) across from side to side. Like the floor of the channel, the inner first stage of the embankment is reinforced by concrete and rises to a height of 13 ft (4 m) at a 40° angle. Wire cages filled with stones form the second stage of the embankment to rise an addition 26 ft (8 m), but at a lesser 20° angle.

Left to nature, runoff descending down through the canyon after a rainstorm would be expected to shoot across the flats to discharge into Puerto Escondido. Confident in their ability to defy nature, humans put into place a fortified installation meant to send water where it normally would not flow. The channel is designed to divert an immense volume of water in a contrary direction. Concrete slabs on the floor of the channel show signs of repair, and those at the outer turn in the embankment bear early signs of distress. One can only imagine the force of water rushing through this structure after

Figure 6.6. Water diversion channel at the base of Tabor Canyon. Photo by author.

a few hours of rainfall related to a passing hurricane. Climbing down into the channel at the mouth of the canyon, we are confronted by what ahead looks like a vast construction site stripped entirely of vegetation and strewn unevenly with coarse gravel. Here and there, naked boulders the size of dump trucks sit partly buried in the gravel. In no time at all, we cross into the canyon at an elevation of 328 ft (100 m), where natural side walls of rust-red andesite draw us within.

The feature that first attracts my attention as a geologist comes into sight at about an elevation of 575 ft (175 m), where the north wall of the canyon towers upward for some 50 ft (15 m) above the streambed. High in the wall, a circular shape roughly 5.75 ft (1.75 m) in diameter marks a lava tube that once torched its way through the underground (figure 6.7). Lava flowing through the tube slowly filled it to the top and eventually solidified in place, leaving a radial cooling pattern. The feature lends a sense of dynamism to what otherwise is

Figure 6.7. Circular cross section of a filled lava tube in the andesite wall of Tabor Canyon (circumference marked by white arrows). Photo by author.

just another wall of immobile rock. Higher in the canyon, a cluster of immense boulders clogs a slot-like channel cut into solid rock. The action of flowing lava caught in a lava tunnel is forcibly replaced by the appearance of these titans that came tumbling down through the canyon flushed by the stupendous force of running water. Recovering from this alternate dynamism, my inner geologist reminds me to examine the boulders more closely for additional bits of information.

The huge boulders may or may not equate with the rocks still held firmly in place by the canyon's side walls at this elevation. Despite their great size, these show the effects of wear, with rough edges having been clipped. If the boulders arrived from far above, their composition would more likely represent a foretaste of the rocks still in place in the canyon's upper strata. Elsewhere clogging the stream-bed, we find igneous boulders oddly mottled in color due to the inclusion of xenoliths, or foreign rocks ripped away from surrounding country rock around subterranean conduits as magma forced its way upward. One need not risk life and limb crawling out on ledges to inspect the same rock carried down as boulders from high above at a level where they remain in place. Indeed, sighting through a pair of binoculars confirms that the peculiar boulders now at our feet trace back to bedrock above us in the upper part of the canyon.

Other details from our immediate surroundings register changes in plant life that are worthy of notice. Above the big boulders, slender trunks of the palo blanco (*Lysiloma candida*) hold fast to the canyon walls. Where this graceful tree is found, one of the few thornless legumes in all of Baja California, water is always close by. A short distance away, a young date palm adds a splash of green to the landscape and confirms the presence of water. It is here that an interval of the sandstone Salto Formation belonging to the Comondú Group makes its appearance. The rocks are just as rust red as the preceding andesite, but reveal a graininess characteristic of sandstone that accumulated in thin layers with traces of cross-bedding. These are among the region's oldest sedimentary rocks, dating back to the Oligocene, possibly 30 million years ago. The unit was first described to the north on the Concepción Peninsula (McFall 1968), where it is said to account for strata 1,000 ft (~300 m) in thickness. The Salto Formation formed largely under subaerial conditions as sand dunes at a time long before the Baja peninsula was pulled away from mainland Mexico.

Climbing upward through Tabor Canyon, the first rock pools are encountered around 1,230 ft (375 m). We pause above what is a great plunge pool (figure 6.8), now holding only a shallow remnant of water from the last rains. It is a sizable basin with plenty of space around its rim. One imagines that Steinbeck and Ricketts stopped here and prepared to camp, while the trackers in the party disappeared higher in search of mountain sheep. None were found during that particular occasion, and the Americans were just as glad for it. During the evening, the locals bantered around a campfire and told funny stories, the punch lines to which the Monterey pair followed the general plot, but failed to grasp with their limited knowledge of Spanish. In the background, tree frogs sang and water striders skimmed over the pond. In so parched a countryside, how did life manage to find its way to this little oasis so far from the next closest body of fresh water? It seemed to them that "life in every form is incipiently everywhere waiting for a chance to take root" at the first sign of rain. In the morning, one of the hunters presented Steinbeck with a handful of sheep droppings as proof positive that the animals inhabited these mountains. In his droll manor, Steinbeck was tempted to accept one of the *borrego* droppings and have it mounted on a hardwood plaque in acknowledgement of an animal not taken, but still alive and healthy when last heard of. The trail steepens

Figure 6.8. Large rock pool in Tabor Canyon with trace of higher water. Photo by author.

dramatically above the next, smaller rock pool. A large boulder the size of a gazebo occupies much of the pool and sits in wait for a coming torrent of water capable of sending it farther down the canyon. I am content with my thoughts and elect to begin my return back through layers of stratified time to the plains below. Stories with secrets aplenty are divulged and the punch line is abundantly clear.

FLIGHT OF FANCY

Commercial airline flights leaving for Loreto from Los Angeles or from Tijuana during the windy winter months almost always descend over the Carmen Passage south of town before banking through an abrupt 180° turn to the north against the wind and in alignment with the runway from that approach. When the wind is brisk and the captain turns the aircraft a little farther south beyond Puerto Escondido, passengers are in for a thrilling spectacle. On such occasions, it is possible to peer down from the left side of the aircraft and view much of the hidden harbor defined in deep blue with a dozen or so sailing vessels at anchor (plate 7). Gleaming white, the sailboats appear like the scatter of large salt crystals on dark icing. The inner hills on the west side of the lagoon retain even proportions at their margins, whereas the asymmetry of the outer hills is evident by comparison. The two barriers that make the hidden harbor what it is, are well defined.

During the same descent, the aircraft flies below the crest of the Sierra de la Giganta, and it is possible to stare into the heart of Tabor Canyon (plate 8). The artificial turn in the altered streambed is easy to follow. Looking back and forth between the mountains and the inner harbor below, the contrast between the natural breakwaters and human-built diversionary channel is startling to contemplate. When will the next storm bring hurricane-strength winds and torrential rain to the district? Will coming storms experienced in this part of the gulf increase in frequency and intensity connected with trends in ongoing global warming? In living memory, the natural barriers have never been known to fail. But I wonder about the artificial barrier at the bottom of the canyon. Is it human folly to so challenge mother nature?

7

Tectonic and Erosional Forces Shaping Isla Danzante

There is a small island next to the mainland [and]
it forms a bay, which has a very good anchorage.
We gave the name Los Danzantes to the bay because
the Indians . . . came out to receive us dancing and
playing flutes made from cane.

Francisco de Ortega, 1633

THE EARLIEST WRITTEN notice of Isla Danzante was recorded by the Spanish mariner Francisco de Ortega prior to the establishment of a permanent mission at nearby Loreto. In time, Danzante became the name associated with the entire island. The original Guaycura inhabitants enjoyed easy access to the island from the peninsular mainland south of Puerto Escondido, which amounts to a separation of only 1.63 miles (2.62 km). What name they bestowed on the place is lost, but Danzante is one of the few Spanish place names anywhere around the Gulf of California that draws a direct connection with the former occupants. A proud non-Spanish name that survives to this day is Mulegé for the river and oasis town located some 75 miles (120 km) northwest of Loreto on Mexico Highway 1 (figure 1.7). Few other names like it are found to exist on the map of Baja California.

Place names taken by towns, rivers, islands, or unusual geographic features tend to be one of three kinds. The most common are patronyms bestowed in honor of a historic personage. The usual practice among Spanish mariners was to name a newly encountered island after one of the many martyred saints, each of whom is celebrated on a particular feast day in the church calendar. More than a dozen islands in the Gulf of California got their Spanish names in this fashion.[1] Others not named after a particular saint still have a religious connotation, like Isla Espíritu Santo (Holy Spirit) or Isla Ángel de la Guarda (Guardian Angel). A few other island names owe their origin to an animal likeness, such as Cabeza de Caballo (Horse's Head) or Tortuga (Turtle). A third category comes under the guise of physical geography, wherein a place is named straight-out for its physical appearance. Isla Pomo translates as Knob Island, and Isla Monserrat refers to that island's jagged uplands. Danzante stands alone as an island name evoking the memory of a people now vanished.

On a personal note, my home state of Massachusetts derives its name from a native tribe, whose members the first colonists from England met on arrival. The natives called their home a place "near the great hill," and we are reminded of their existence by a state flag bearing the semblance of an American Indian holding a bow and arrow. Marking the boundary between western Massachusetts and New York State, the Taconic Range is visible from my house. It forms a distinct section of the worn Appalachian Mountains, deriving its name from the native word Taghkanic for a place "in the trees" as well as the name of a historic chieftain. Place names possess a fascinating potential to be transmogrified into something having a new and overriding meaning. For geologists, the Taconic Range is synonymous with a major tectonic event that resulted in the growth of Andean-style mountains along the margin of the ancestral North American continent.[2] With a name so evocative of rhythmic movement, Danzante Island is a fitting choice for a repurposed metaphor attuned to the geologic origins of gulf islands and the cadences by which they are sculpted over time. Based on the beauty of its natural features and easy accessibility, there could be no better choice.

FEATURED EVENT
Visit to Isla Danzante

Ranked as one of the smaller islands in the Gulf of California, the slender island of the dancers is no more than 1.75 square miles (4.64 km²) in size, but reaches a central height of more than 1,180 ft (360

m) above sea level (map 7). Together with Isla del Carmen (see chapter 5), Danzante is one of several islands under conservation protection within the Loreto Bay Marine National Park. Visitors are prohibited from collecting rocks or disturbing the island's plants and animals. A private tour by small boat with stops along the way to hike its scenic trails may be arranged with a departure from the ramp at Puerto Escondido (see chapter 6). Including transportation from and returning to Loreto, a full day's exploration entails a leisurely coastal circumnavigation amounting to 9 miles (14.5 km), with landings along the western shore to access two short trails collectively no more than three-quarter's mile (1,200 m) in length. What a visit to Danzante lacks as a physical challenge, it more than

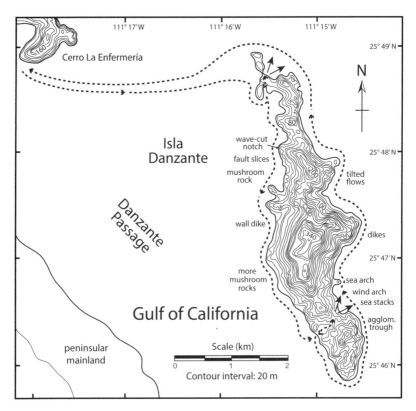

Map 7. Topographic map of Isla Danzante showing the tour route from Puerto Escondido.

compensates for in terms of exposure to concepts of gulf tectonics and coastal erosion.

From Puerto Escondido, the morning's trip across to the north end of Isla Danzante covers 2.75 miles (~4.5 km) and requires only a few minutes travel on a calm sea (map 7). Our immediate destination is the very area where Ortega encountered Guaycura natives in 1633, now rechristened for today's tourist market as "Honeymoon Bay." A few random cardón cacti (*Pachycercus pringlei*) stand like sentinels at stiff attention, but with tenuous footholds on the west-facing bluff. They provide a useful scale against which an initial impression of the rock formations at the opening to the inlet may be measured. Like nearby Puerto Escondido, the geology here is dominated by andesite belonging to the Miocene Comondú Group.

Whereas rocks in the seaward hills around the harbor are relatively thin layered and steeply tilted due to tectonic deformation subsequent to their origin as volcanic flows, the bluffs that shelter the inlet of the dancers are more massive and betray no hint of inclination. A lower band of gray rocks rises 9 ft (~2.75 m) from the waterline, overlain by more deeply weathered bands of rust-red andesite that add another 50 ft (15 m) in height. The red rocks exhibit deep furrows that look something like the folds in an accordion. On closer examination, the creases pass vertically downward through the less weathered rocks below. They represent the natural joints formed in lava flows as the molten material began to cool and contract. It appears that two or perhaps three separate lava flows flooded the original topography long before the island was separated from the mainland. The lower, gray band of rock exhibits less exposure to the elements and is likely to have risen above the waterline during the last 125,000 years. Rounding the corner of the bluff, the boatman eases on the engine's throttle and we glide toward the beach at the far end of the inlet. It is one of only two sandy beaches anywhere on the island, and it stretches out for a generous distance of 160 ft (~50 m) from one side to the other at the termination of a flooded box canyon.

At the east end of the beach, a well-marked trail leads upward to any number of vantage points where the surrounding topography may be studied. The first to attract the geologist's attention is the

view out over the inlet we've entered (figure 7.1). Details suggested
on the first approach to the island are confirmed, in particular the
vertical jointing and the level posture of successive andesite flows.
Continuing higher to another lookout oriented toward the open
gulf, a deep cleft in the rocks cuts downward to the seashore some
100 ft (~30 m) below. It is possible to descend into the defile, which
has the width a little greater than outstretched arms to either side.
From within the cut, the blue sea cuts a v-shaped punctuation
mark in the rocks, with the volcanic cone of Isla Coronados on the
horizon, 20 miles (32 km) in the distance. The cleft's alignment is
only a few compass degrees west of due north. Higher on the east
face than the west, the cut has every appearance of an entrenched
fault scarp. The orientation differs from the parallel fault traces
along the inner and outer margins at Puerto Escondido (compare
with map 6).

Figure 7.1. View overlooking the cove and beach at the north end of Isla
Danzante. Photo by author.

On a side trail to the north, another vantage point opens onto the far tip of Isla Danzante (plate 9). It is apparent that the distal end is entirely cut off from the rest of the island during high tide. The two pieces of rocky real estate are connected by a cobble beach in the form of a tombolo. The view offers another look at the distinctive vertical fractures that cut through thick andesite flows stacked in succession like so many layers in a tiered cake. What occurred here was a violent breach through a formerly solid rock wall, weakened by fractures where the wall was least fortified. The distance separating the islet from the rest of Danzante is not so great, but one can make a comparison with the islet at Puerto Escondido and the boulder barriers linking it to hills on opposite flanks (see chapter 6). The tides come and go with a regular periodicity day in and day out, but the extra energy responsible for the ultimate breakthrough was dispensed by big storms that pounded the island on an episodic basis from the open sea to the east.

On our return to the beach, the trail passes a lone torote copal (*Bursera hindsiana*) prostrated by the wind. It is a fine example of krummholz (meaning twisted wood), deformed by wind-borne salt aerosols.[3] The lone tree is like the wind sock at an airport that indicates wind direction. In the case of the tree, the wind has arrived seasonally but always from the same direction blowing north to south. In the space of merely an hour, we have been exposed to a range of physical parameters that shaped Isla Danzante: rocks with inherent internal weaknesses, faults, tides, storms, and seasonal winds. All such factors are significant over time, but how can we tease apart their efficacy as nature's sculpting tools in the lifetime of a single island? Confidence in reading the rocks increases with the recognition of recurrent patterns.

The next observations are viewed from the boat about two-thirds of a mile (1 km) south of the box canyon with its ample beach. Motoring slowly close in to shore, the island's ridge line is seen to rise in height, with few intervening lows from around 260 ft (80 m) to an elevation in excess of 655 ft (200 m). A bulbous knob of rock forms a prominent headland tracing inland to the center of the island. The first sign of something out of the ordinary is the overhang that projects above the water. It is a wave-cut notch (map 7) eroded at the base of the headland within the intertidal zone. The size of the notch is impressive, due in part to the range of daily changes in the tide line. Here, the roof of the overhang easily reaches 8 ft (2.5 m) above the waterline at low tide. It is not the tide itself that eats into the rock, but attendant wave action in conjunction with the tides. The critical factor that either impedes or enhances the rate of erosion is the hardness of the rock. Looking back to the north after passing the headland, the indentation of the notch is more readily apparent (figure 7.2).

Other aspects of geologic significance are viewed from the south side of the headland. In particular, its profile is broken into no less than four conspicuous shapes that rise in elevation and become increasingly more pointed inland. In Nordic countries, folklore posits that unusual shapes in mountain rocks are trolls caught in the morning daylight after a night of play. If unable to safely reenter the mountain before the first rays of sunlight at dawn, they are locked

Figure 7.2. Line of hills on the west side of Isla Danzante offset by parallel faults that sliced through a previously existing ridge to demarcate separate peaks. A prominent wave-cut notch is incised at the base of the hill nearest the shore. Photo by author.

outside and forever captured in solid rock. It is tempting to invoke the image of four large, clumsy trolls emerging from a nighttime swim and attempting to clamber back into the heart of Danzante Island before daybreak. But geomorphology offers salient explanations as to why mountain formations take on the shapes they possess. Based on the size of the saddles eroded between adjacent peaks and the descending v-shaped drainages that fall below, the pattern on an east-west oriented ridge is supported by a set of parallel faults that cut across with a roughly north-south orientation (figure 7.2). Such trends are reminiscent of the big fault trace previously observed from the trail at the island's north end. Another detail of geologic importance is that andesite layers forming the ridge are not only well defined, but reflect a distinct dip of roughly 30° to the west. In contrast, the massive andesite layers around the box canyon (figure 7.1) and at the island's north end (plate 9) are flat.

Farther along, the next phenomenon expands on the number of interrelated geologic puzzles belonging to Isla Danzante. The feature is a rock pediment, also called a mushroom rock or hoodoo (figure 7.3). It is one of the most celebrated peculiarities on the island, and well-known to park guides. A similar example at Balandra Beach near La Paz is more famous, but its pedestal broke and caused such an outcry that it was artificially restored in 2005. Fewer visitors come to Isla Danzante, and the shore on which the pediment sits is off limits for landings. Balanced on a pedestal little more than 18 in (45 cm) in diameter, the 80-in (2-m) high structure is fragile. There is, however, an enhanced beauty to the Danzante mushroom rock and others in the process of development nearby, because their natural development can be readily understood. Basically, the mushroom cap is composed of harder volcanic stone than the undergirding neck.

Quite another example of differential erosion that occurs entirely ashore is met about a third of a mile (0.5 km) farther south. At first it looks like an enterprising rancher built a magnificent stone wall that marches across a valley and up the hillside to the base of a high rock cliff. But the structure is not man-made. It is an exhumed geologic dike (figure 7.4), part of which has been breached by erosion. The height of the standing wall is astonishing, vertically at least 25 ft (7.5 m) or more. Full grown cardón cacti nearby are overshadowed by

Figure 7.3. A rock pediment (mushroom rock) eroded by waves and currents within the tidal zone. Photo by author.

comparison. Layered andesite crops out on opposite sides of the dike, although overgrowth of desert brush obscures much of the hillside. The igneous rock that forms the dike is harder than the volcanic layers it passes through. Sea level never rose to where the wall now stands. The softer volcanic rocks parted by the dike simply were deflated by wind erosion. Over thousands of years, the seasonal winter winds that blow down from the north were adequate to the task.

Another mushroom rock occurs at the shore in the same area, and the mystery of its formation is more evident. At low tide, adjacent sea cliffs are found to be composed of a kind of volcanic conglomerate that goes by the technical name "agglomerate." Whereas a sedimentary conglomerate is formed by a collection of eroded clasts that

Figure 7.4. Igneous dike crosscutting through layered andesite flows and now left as an exposed vertical wall due to differential erosion by wind and rain. Photo by author.

become cemented to one another, an agglomerate is derived from a mass of volcanic rubble that moves as a flow when hot. On cooling, the agglomerate solidifies and is held together in a matrix of finer-grained volcanic material. From the boat, it is possible to view a line of pediment rocks half exposed as a kind of frieze in the cliff face (figure 7.5). Each structure is separated from its neighbor by a vertical fracture. Small excavations appear in the lower part of the tidal zone, where the softer matrix of the agglomerate is subject to intervals of wave or current erosion. In profile, the emergent rock pedicles look like a row of solid cones turned upright on their toes. An isolated mushroom rock occurs when its closest neighbors on opposite sides lose their pediments and the cap rock collapses between defining vertical joints. When all the rock pediments are excised by erosion, something like an extended wave-cut notch may survive, as viewed earlier (figure 7.2). The layers that dip into the sea from the big headland are, in fact, built up from successive agglomerate flows, some of which are denser than others.

The Danzante Passage separating Isla Danzante from the peninsular mainland is not very wide and is sheltered to the north by Cerro La Enfermería at Puerto Escondido (map 7). But the channel is open to longshore currents stimulated by the winter winds that sweep across the land and push ocean swells alongside. Development of mushroom rocks on the western side of the island is more likely due to currents that brush parallel to shore than to storm waves striking from a perpendicular direction. The energy of even a moderate storm wave would be more than enough to break the pediment of Danzante's star mushroom rock.

Figure 7.5. Coastal bluff on the west shore of Isla Danzante that prefigures erosional stages leading to the development of freestanding rock pediments. Photo by author.

The second-largest beach on Isla Danzante is found on the west shore below a prominent saddle in the island's north-south axis. From the landing place, a well-marked path leads to a viewing point on the ridge crest (map 7). The authorized park trail ascends to an elevation of 280 ft (~86 m) and is a half mile (800 m) long. As the overlook from the saddle is special and deserves adequate time to appreciate, we bring with us our lunch for a midday pause. Behind the beach, the dominant ground cover is the mangle dulce (*Maytenus phyllanthoides*), a shrub with fleshy leaves that does especially well in saline soil. In short order, the mangle dulce is replaced inland by the palmer mesquite (*Prosopis palmeri*), also prone to form thickets. In turn, the mesquite is replaced at a higher elevation by a mixture of palo adán (*Fouquieria diguetii*) and torote copal (*Bursera hindsiana*). Closer to the saddle and especially against the valley margins, palo blanco trees (*Lysiloma candida*) are more common. Here on the west shore, the basin through which we climb holds its moisture and promotes a rich diversity in vegetation otherwise absent on steeper, rocky slopes.

After a final climb through step-like layers of naked andesite, the trail ends overlooking a bay open to the northeast. The change in geomorphology from the valley is abrupt, and we are confronted by shear sea cliffs accentuated by layered andesite. The reward is a dioramic view over the coast that includes sea stacks and "The Window"—a sea arch with a remarkably square opening cut through a solid spur of rock thrust out like an arm to the east (figure 7.6). As our plan is to complete a circumnavigation of the island, we will have the opportunity to investigate the arch at closer range from the water. Under repeated but not necessarily constant wave attack, a rocky shoreline is prone to excavation of sea caves. A sea arch is born when two sea caves intersect from opposite sides of a rocky spur. Moreover, a sea stack results from ongoing erosion against the inner walls of a sea arch such that the span of the arch is widened until it collapses under its own weight.

The most celebrated place on the Baja California peninsula where the progressive evolution of sea caves into sea arches and sea stacks may be witnessed is Land's End at Cabo San Lucas (figure 1.7). There, the rocky shoreline is governed by massive granite. Here on

Figure 7.6. Small sea stack (lower left) and sea arch (center) eroded in andesite layers on the east coast of Isla Danzante as viewed from the lookout point at the end of the south trail. Photo by author.

Isla Danzante, the coast is dominated by andesite but with subtle variations in layering and textural composition. The rock type is not so critical as the presence of fractures susceptible to hydraulic hammering through wave impact. The bay that stretches out before us with an area of approximately 25 acres (~10 ha) has been carved through the gradual erosion of sea caves, sea arches, and sea stacks. Taking into account the precipitous sea cliffs that line the bay, the amount of rock removed by such action is impressive. The effect of time and erosion on the more exposed shore is worth contemplation while enjoying a lunch break.

Returning to the beach where our boat awaits, further thought is warranted regarding the preferential occurrence of a sandy shore on the west side of the island. What does it mean if the mushroom rocks also occur only along the same shore? The distance navigating around the south end of the island to reach the embayment viewed from the trail above amounts to a mile and half (2.4 km). Variations in layering and orientation of volcanic flows that make up the island's rock core take on added significance, given the differences between sheltered and exposed shores. Volcanic flows may pond on a flat surface, but they also are subject to the pull of gravity on the least-inclined slopes. Heretofore, andesite beds have been observed flat-lying or tilted to the west where encountered along the Danzante Passage (compare figures 7.1 and 7.2). Tectonic forces surely played a role in post-depositional tilting. Is there any sign of a preexisting landscape that may have influenced flow direction at any given time?

Such evidence is preserved in the two-dimensional wall of the outer coast just short of the embayment with its three-dimensional sea stacks and sea arch. There are so many competing attractions to balance during a visit to a desert island in the Gulf of California. Others may pass this spot with little notice, but I am transfixed by a sight that tells something about the lay of the land more than 10 million years ago. A cardón cactus provides the needed scale against which to appraise a succession of events that transpired at this place long ago (figure 7.7). The first was a volcanic flow that ponded on a flat surface, but after hardening as andesite was dissected by surface erosion to form a trough during a pause in volcanic activity. The second event brought another volcanic flow that filled the trough to a depth of 26 ft (8 m) until it nearly topped out. After yet another pause in volcanic activity, a coarse agglomerate buried the andesite in the trough. The contact between the upper two units is sharp and follows a slight depression tracing the last remaining contours of the surface. It is critical to remind ourselves that Isla Danzante did not exist when these events occurred. In fact, the Gulf of California in which the island now resides was yet to appear. Even so, this sliver of a small island retains story fragments that occurred long before the tectonic events that opened the Gulf of California and led to further contortions of those volcanic layers.

volcanic flow
filling a trough

Figure 7.7. Rock wall on the lower east side of Isla Danzante that illustrates a sequence of events including a volcanic flow that filled a trough-like depression in the former landscape (cardón cactus for scale). Photo by author.

Northward, we soon arrive at the embayment where the first and most impressive of two sea stacks comes into view (map 7). The tide has yet to reach its nadir. The sea stack is exposed in the middle of a narrow beach, where it sits on a square base with horizontal bedding that tapers gracefully upward to a pinnacle at an elevation maybe 60 ft (~18 m) above the water level. Another, much smaller sea stack sits a bit farther along, encircled by an oval-shaped tidal flat covered by stony rubble. In fear of missing important details, I motion to the boatman to reduce our speed to a crawl while I scan the cliffs behind the sea stacks. Briefly and high above, a wind-carved arch is spotted at the top of a ravine. The upland arch was not visible from the end of the trail, earlier. It makes a fleeting appearance as an unexpected bonus and testifies to the vigor of the seasonal winds capable of digging into pockets of softer rock.

Full ahead, we drift toward the island's prize monument: "The Window." Maybe 25 ft (~7.5 m) in height, the roof of the structure is formed by a lintel bridging an open space. The crosspiece is, in fact, a single stratum of volcanic rock that supports an impossibly thick body of strata above. Given its profile, this is no ordinary sea arch. Many large stones that have fallen from the roof now litter the tidal flat beneath. When lifted by a rising tide, it would be possible for our boat to cruise through the opening and emerge on the other side with room to spare. As we approach closer, a vertical fault is seen at one side of the lintel that runs through the top of the roof (plate 10). The east side of the structure has settled downward a few feet (~1.5 m) relative to the rest. Side-wall erosion has widened the window, but the critical factor in its enlargement is roof collapse due to structural weakness introduced by the fault. The lintel is not only a wonder to behold, lending insight to the window's origin, but also a fatal flaw that portends the structure's future demise. It may require hundreds of years, but ongoing wear and the pull of gravity will someday transform the arch into a sea stack. Such a prediction does not detract from a realization that the splendid sea stacks now out of sight behind us were once connected by a bridge of stone to the rest of the island.

Clinging to the coast on a northeast heading, the next stage in our transit brings us past the widest part of Isla Danzante, less than a mile (1.6 km) in girth. Checking the topography (map 7), two things are confirmed. First, the island's maximum width also features its highest peak at an elevation exceeding 1,115 ft (340 m) above sea level. Second, the cross-sectional profile along this line is surprisingly symmetrical. That is to say, the opposite slopes on the east and west sides of the island are much the same. The extreme erosion on the outer flanks of hills around Puerto Escondido (chapter 6) is not apparent in this section Isla Danzante. It may be argued that coastal attrition is somewhat more active on the island's outer flank, but not excessively so. Vertical dikes make their appearance along this part of the coast. The great wall dike viewed earlier in the day above the west shore (figure 7.4) is sure to match a dike on the east shore. It would not be unusual for such a feature to cross clean through the island from one side to the other. In contrast, none of the dikes here exhibit the same degree of exhumation as observed on the opposite shore. Andesite flows dissected by crosscutting dikes along this part of the coast are flat in two dimensions, much the same as the other side of the island at the same latitude.

The disposition of andesite flows changes as we continue. Individual layers exposed in the sea cliffs here are now inclined roughly 20° northward (map 7). To be sure, the inclination registered in sea cliffs viewed straight on from the east is an apparent dip. The true dip could be much steeper and more angled to the northwest (or southeast). On the opposite side of the island at a similar position, we saw the march of the trolls from different angles. Separated from one another by parallel faults (figure 7.2), each of the rocky knobs was seen to be formed by andesite layers with consistent inclinations to the west or northwest. From limited observations like these, it might be surmised that some parts of the island are more tectonically impacted than others. At the close of the Danzante circuit, several extraneous but oddly related lines of thought converge as we begin our run back toward Puerto Escondido (map 7).

CLOSING IMPRESSIONS ON ISLAND RHYTHMS

Why are mushroom rocks found only on the island's west shore? Why do sea stacks and the dramatic "Window" occur as features only on the east shore? We can be confident that some topographic variation existed during the span of time when individual volcanic flows overlapped one after the other in this area. To the extent visible, many volcanic layers ponded and cooled to form level accumulations one atop the other. But some flows poured downslope to fill trough-shaped depressions in the original landscape. The peculiar circumstances that led to the formation of pediment rocks relied on a fortuitous alignment between softer and harder rocks at just the right exposure within the contemporary tidal zone. Such conditions might occur on the east shore, but any resulting structures are more prone to total destruction from storm waves. Longshore currents brush both shores on a seasonal basis, but the deft touch of such sculpturing is more likely to persist in the relative shelter of the Danzante Passage.

Like many other islands in the Gulf of California, Isla Danzante is a fault-block structure left standing above the waterline as a horst. Lesser faults appear to run mainly north-south through the island, but diagonal faults could also be present as more accurately mapped on nearby Isla Monserrat.[4] In contrast, the Danzante Passage is a half graben. Formerly at the surface, the adjacent fault block sank beneath the waterline between parallel north-south oriented faults. Those faults were activated prior to about 3.5 million years ago, as the earth's crust was subjected to extensional forces pulling in opposite directions from below. The average water depth in the Danzante Passage is 118 ft (36 m). The island itself is bound by roughly parallel north-south faults. A more deeply submerged graben is aligned to the east, where the water depth drops off rapidly to 145 fathoms (265 m).

Compared with the landscape surrounding Puerto Escondido (map 6), the defining faults trace a northwest to southeast orientation parallel to the master Loreto fault and more consistent with the later and ongoing phase of tectonic activity in the Gulf of California initiated about 3.5 million years ago. Individual gulf islands need not be extensively dissected by such faults, but will slide laterally together with larger segments of the surrounding seabed in concert with transtensional tectonics powered by seafloor spreading within the gulf's deepest basins. It is a topic treated in greater detail in the following chapter on Santa Cruz and San Diego islands. The base beat of ongoing gulf tectonics is overprinted by the syncopation of seasonal patterns in coastal erosion predicated by wind, rain, and major storms. Isla Danzante is a practical and ideal place to sense those rhythms.

8

Reaching for Islas Santa Cruz and San Diego

The Gulf of California is a young ocean basin
formed by oblique extensional motion between
the North American and Pacific plates and is
an example of a common class of rifts.

*M. Lyle and G. E. Ness (1991) in The Gulf and Peninsular
Province of the Californias*

THE SEA OF CORTEZ resides in a youthful ocean basin, but it is a strapping offspring already 100 miles (161 km) wide at a latitude north 25°14', passing between Islas Santa Cruz and San Diego. These small islands are located 12.5 miles (20 km) offshore of Baja California, separated north to south by only 4 miles (6.5 km). Physically, the islands sit on the edge of the peninsular shelf near where it drops off to a depth of 3,000 ft (915 m). A little more than halfway across the gulf eastward from Isla Santa Cruz, the sea reaches an astonishing depth of 20,000 ft (~6,100 m). Given the Earth's curvature, line of sight on a clear day from a small vessel is little more than 3 miles (5 km). The peninsular mainland recedes from view well before reaching those particular islands on a trip from the closest departure point. The gulf's expanse and its extraordinary depth contribute to a sensation of smallness, especially when visitation is by means of an open fisherman's panga.

Geology lends another layer of perception to any journey in hope of reaching Isla Santa Cruz and its neighbor Isla San Diego. Unlike Isla del Carmen (chapter 5) or Isla Danzante (chapter 7), both of which require short commutes from peninsular shores, Santa Cruz and San Diego are different and yet surprisingly akin to the axial core of Baja California. Carmen and Danzante are formed by Miocene andesite younger than 23 million years, whereas Santa Cruz and San Diego are composed of granite dating back to the Cretaceous Period, roughly 84 million years ago. The same granite is exposed along the spine of Baja California, where it gives a surreal touch to the landscape as experienced around Cataviña[1] in the north or to the mountains above Cabo San Lucas in the south. Granite is the product of magma that cools slowly in chambers far below the earth's surface. Only when the overlying layers of rock that insulate those chambers are unroofed by erosion is the granite exposed to the light of day long after magma turned to solid stone. Andesite and sandstone from the Comondú Group contribute to the thick overburden that more often conceals the granite, as encountered at Tabor Canyon (see chapter 6). On uninhabited islands like Santa Cruz and San Diego where landings are difficult, part of the allure is their nakedness stripped of the Comondú rocks that covered everything prior to the tectonic events separating Baja California from mainland Mexico.

Few other islands in the Gulf of California are formed exclusively by granite.[2] The largest among them is Santa Catalina, the outlying island in Loreto Bay Marine National Park. Tourist vessels that cater to ecotourists on week-long excursions in the Gulf of California sometimes stop at Santa Catalina, which offers a suitable alternative for a visit to a granite island on account of access to valleys leading at a low incline to the summit. I have visited the larger island, but I am drawn to the smaller islands out of an innate stubbornness in favor of entities with better clarity in proportion to their size. Remote and small as they are, the Santa Cruz and San Diego islands are enticing for the stark insights they offer on the Gulf of California and its geologic history.

FEATURE EVENT

A Run at Islas Santa Cruz and San Diego

South of Loreto, Mexico Highway 1 makes a detour far inland away from the gulf coast with few secondary roads that offer access to the region until nearly as far as La Paz (figure 1.7). Departing from Puerto Escondido, the sea route to Isla Santa Cruz requires a day's leisurely travel by motorboat. Because the island offers precious little in the way of a suitable camping place, a stopping point on the peninsular coast is necessary and the beach at El Gato makes a convenient choice. Several places in Baja California take "The Cat" as a place name, but there is only one embayment under that colloquial name, and it is located 36.5 nautical miles (68 km) southeast of Puerto Escondido. I spent an enchanted night camping at El Gato with colleagues and an experienced guide on my way down the peninsular coast during an epic, two-week tour between Isla del Carmen and Isla San José in January 2003. Especially with regard to geology, the stop offers a vital point of transition on the way out to the granite islands.

Only a limited exposure to the Salto Formation within the greater Comondú Group dominated by andesite and tuff is available at Tabor Canyon (see chapter 6). But at El Gato, the sandstone so typical of the formation crops out more generously at sea level. Like plump pillows aligned with the contours of cross-bedding, the sandstone seems to billow forth in a soft breeze. Formed during Oligocene time, the sandstone was emplaced long before the separation of the Baja California peninsula from the Mexican mainland. Cross-bedding is diagnostic for sand dunes that blew into place through dry valleys, although some layers with planar bedding also are interpreted as lake-bottom sands. Rust red by nature, the rocks glow deep red in color and cast long shadows in the setting sun. The relationship to the much older Cretaceous granite is significant, because the sandstone represents the basal part of the Comondú Group that came to bury an older granite landscape.

The topography of Isla Santa Cruz (map 8) is confined to an island barely 5 square miles (13 km^2) in area, but with a peak elevation 1,700 ft (520 m) above sea level. In map view, the outline looks something like a curved bow with a taunt bowstring. Noticeable at once from variations in topography, the granite island is asymmetrical, crossing a north-south axis with the steepest terrain rising in cliffs from the east side. From the west shore, a dozen deeply entrenched arroyos claw inland to reach an elevation 1,300 ft (~400 m) above sea level.

The longest is Cañada La Leña, which extends eastward for two-thirds of a mile (1 km). Tracing line A–A' inscribed between Cañada La Leña and arroyo La Crucecita (map 8), a typical cross-sectional profile demonstrates the island's asymmetry (figure 8.1). In this treatment, the vertical scale is exaggerated over the horizontal scale by a factor of three to one, making the contrast in slopes on opposite sides of the island more obvious.

The scenario for Isla Santa Cruz is similar to that observed previously when visiting the islet linking the two natural breakwaters at Puerto Escondido (review figure 6.5). In that example, the exposed

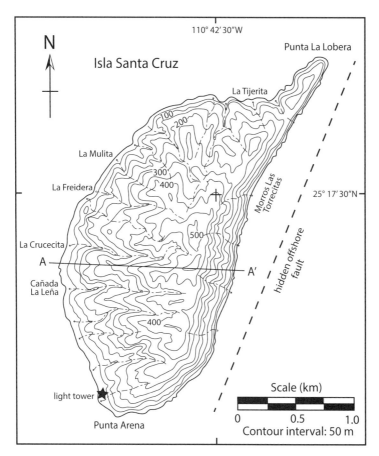

Map 8. Topography of Isla Santa Cruz demonstrating the island's severe asymmetry on a roughly north-south axis.

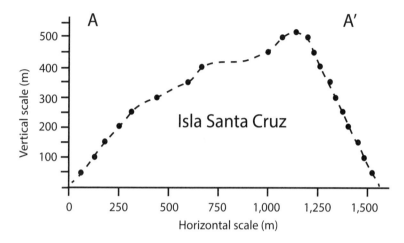

Figure 8.1. Cross-sectional profile through Isla Santa following the line A–A' in map 8 halfway between Cañada La Leña and arroyo La Crucecita. Vertical exaggeration: 3x.

outer side of the islet was found to be severely eroded along a fault scarp. Much the same exercise in geomorphology can be applied to Isla Santa Cruz, wherein a more symmetrical island may be reconfigured by redrawing lines of topographic elevation to mirror the same spacing on the east side as presently exist on the west side. As much as a third of the island's original mass may have been removed by preferential erosion on the eastern flank. Like Isla Danzante (see chapter 7), Santa Cruz is the product of tectonic forces that generated a small fault block left above sea level as a structural horst. Structural grabens sit parallel to the island below the water line on its western and eastern flanks. Because Santa Cruz is farther away from the peninsular mainland than Danzante, the drop off in water depth is greater on both sides, but substantially more so on the eastern side. Notably, the fault trend on the east side of Santa Cruz follows an orientation 22° east of north, which is unlike the orientation of faults observed at Isla Danzante (north-south) or Puerto Escondido (northwest to southeast). Hence the local history of tectonics is different here at Santa Cruz, and we are challenged to understand why that is the case.

It is an accomplishment simply to have reached Isla Santa Cruz on a calm day in anticipation of a leisurely circumnavigation of the island at close range. From a geologic point of view, the task at hand is to experience the place from all sides while following along with the topographic map. Off the northeast end at Punta La Lobera (map 8), the cragginess of the rocky coast is blunt and utterly barren (figure 8.2). The point follows back along the island's principal topographic axis for a third of a mile (~530 m) and rises precipitously to an elevation in excess of 500 ft (152 m) above sea level. Natural weathering at Punta La Lobera suggests a pattern of exfoliation in the granite, with thick intervals dipping off to the west or north west. As commanding as they are, the sea cliffs farther south along the island's east shore attain greater heights, as much as 1,640 ft (500 m) at a place where the map gives the name Morro Las Torrecitas (map 8).

The colorful name might be loosely translated into English as "Snout of the Little Towers." Here, the rocky shore rises vertically for at least 165 ft (~50 m) and thereafter assumes a gradient equal to a 45° slope (figure 8.1). Viewed close-in to shore (figure 8.3), the high wall of granite is scored with fractures that cross one another with an even regularity. Few plants find a foothold on these cliffs. At low

Figure 8.2. North end of Isla Santa Cruz at Punta La Lobera. Photo by Hank Ellwood.

Figure 8.3. Vertical rocky shore of Isla Santa Cruz along Morros Las Torrecitas on the east shore, showing fractured granite and a thin trace of a gravel
beach at low tide. Photo by Hank Ellwood.

tide, only the narrowest of a gravel beach appears above the waterline. Yet farther along (plate 11), the steep granite coast reveals more
signs of exfoliation when viewed from a wide angle that includes the
island's south end in the distance. Here, the granite is shot through
with white bands of quartz. The hard quartz accumulated through
remobilization of silica that penetrated the granite after the cooling
process had advanced much closer to completion. The island's only
sand beach is found at Punta Arena, near the light tower (map 8).
The coarse sand is a quartz derivative and on closer approach appears
bereft of any significant shell content.

Continuing the tour, a change in island slope along the west shore
is obvious where framed by a series of arroyos with open mouths
gaping to the sea. We may peer into the longest canyon through
the entrance to Cañada La Leña (map 8), although neighboring

Arroyo La Crucecita (The Little Cross) qualifies as a close second. Here between the two valleys, the cross section through Santa Cruz achieves an average 12° slope, which is equivalent to a 21 percent gradient across the lower two-thirds of the rise (figure 8.1). Remnants of a former but very narrow plateau crown the island, broken into hillocks rising some 165 ft (~50 m) above an undulating ridgeline (map 8). Few humans have attempted to thread a way up into the arroyos to reach those hillocks. The east coast would stop all but the most dedicated enthusiasts for rock climbing with safety ropes, but the western slope is not much easier. On foot, a climb through a 20 percent gradient could be treacherous. The pull of gravity with a misstep has its consequences. It is useful to be reminded that a roadway with a 10 percent grade is sufficiently steep to warrant a road sign urging caution to check the brakes.

Farther along the western shore, the tedium of granite is broken by the appearance of sedimentary rock unmistakable as conglomerate (figure 8.4). The conglomerate forms a distinct unconformity, perched as it is atop a cliff of granite 33 ft (10 m) above sea level. Conglomerate materials consist of granite that was eroded by wave action from the parent basement rocks. Well-rounded stones range in size between cobbles and small boulders, judging by the size of a nearby cardón cactus rooted in gravel. The deposit is thoroughly cemented, but open pockets can be spied where individual boulders were loosened and fell to the shore below. Given the deposit's lateral continuity and associated unconformity, a former wave-cut shelf is implied that underwent uplift to form a marine terrace. As noted previously (see chapter 5, plate 5), marine terraces throughout the Gulf of California are interpreted as the product of changing sea level in tandem with coastal uplift by tectonic means. The singular example of a marine terrace at this elevation on the west side of Isla Santa Cruz is likely to be late Pleistocene in age (roughly 125,000 years old). Global sea level was as much as 20 ft (6 m) higher during the last interglacial epoch than today, due to extensive melting of landlocked ice at higher latitudes.[3] But Santa Cruz and other islands like Carmen also were subjected to gradual uplift by as much as 20 ft (6 m) in subsequent years due to the play of ongoing tectonics in the Gulf of California.[4]

Figure 8.4. Rocky shore of Isla Santa Cruz along the west side, showing contact between granite basement rocks and overlying conglomerate representing a former beach now elevated well above present-day sea level. Photo by Hank Ellwood.

A better handle on the Santa Cruz scenario requires a boat landing and a risky climb to reach the conglomerate in search of fossils. Close to the granite shores at Punta San Antonio in the midsection of the gulf to the north (figure 1.7), such a study was undertaken to describe Pleistocene fossils encrusted and preserved in life position on eroded boulders in a similarly thick conglomerate (Johnson and Ledesma-Vázquez 1999). The same sort of study would be invaluable here, but logistically hard to pull off. Even at a rudimentary stage of investigation, however, the following facts pique our curiosity. There exists an obvious contrast on a small island between a west-facing, low-angled slope with an etched marine terrace as opposed to an east-facing, over-towering cliff face lacking any sign of a marine terrace. The rise and fall of sea level is a common occurrence through geologic time. Tectonic blocks are vertically hoisted as structural horsts or sunk as grabens. Through these dual processes, a marine terrace might be expected to survive like the brim of a hat around all sides of a fault-block island. Assuming such a terrace once existed on the east side of Santa Cruz, it has now vanished through the agency of asymmetric erosion on one flank compared to the other.

How does life reach a small and remote island like Santa Cruz to initiate a founder population? For intertidal mollusks like the bivalves (*Chione californiensis*) and sea snails (*Turbo fluctuosus*) found as fossils in the hills above Puerto Escondido or among the conglomeratic boulders near Punta San Antonio, the challenge of migration to a fresh habitat was not difficult. Reproductive spawn (eggs and sperm) are shed directly into seawater by mollusks and most other marine invertebrates on a seasonal basis, where they combine in the act of fertilization to form a free-swimming propagule subject to transport by the tides and currents. Barnacle propagules, for example, mature quickly and will expire if attachment to a suitable rock or shell is not found after a short time. Other invertebrates can "tread water" for a longer period of time, during which opportunities for reaching a safe haven are multiplied. A mollusk propagule is capable of wiggling about on its own account, but passive drifting within a current of moving surface water is what allows marine invertebrates to more quickly migrate from one place to another.

Most fail, because they are taken as food by other organisms near the base of the food chain. But a few succeed to renew the reproductive cycle in another place. By this means, it is possible for marine invertebrates to jump from one island shore to another through multiple generations over the length and breadth of an expanse like the Sea of Cortez. Indeed, some colonists are thought to have reached the Gulf of California from places far away. For example, a fossil stony coral (*Solenastrea fairbanksi*) commonly found at Punta Chivato has its closest living relatives in the Caribbean Sea on the opposite side of the continent. It is assumed to have made its way as a propagule through an open passage separating North and South America before the Panama Isthmus became a permanent barrier about 3.5 million years ago.[5] Likewise, a fossil bryozoan (*Conopeum commensale*) discovered in Arroyo Arce north of Loreto has its nearest kin also in the Caribbean (Johnson and Cuffey 1997), and is regarded as another far-traveled colonist. It appears that both the coral and bryozoan went locally extinct in the Gulf of California, leaving no descendants. A proper survey of marine invertebrates living in the intertidal zone around Isla Santa Cruz has yet to be made.

Plants constitute quite a different challenge with respect to island colonization. Birds that consume the fruits of desert plants may carry seeds from place to place, which are left behind with droppings. Seeds also may be transferred in the mud adhering to the feet of wading birds, and hence carried from island to island. Vegetation growing above the conglomerate bed on Santa Cruz (figure 8.4) includes the cardón cactus (*Pachycereus pringlei*), the coast cholla (*Cylindropuntia prolifera*) and brushy palmer mesquite (*Prosopis palmeri*), all of which provide only thin cover. A scattering of cardón also are visible on the island's steep eastern flank (plate 11). Bats are known to visit cardón cacti at night, and they are the necessary agents carrying cardón pollen from plant to plant for cross-fertilization.

Reptiles are more problematic with regard to island colonization. Probably the most extensive field study with respect to island biogeography in the Sea of Cortez has been conducted on reptiles.[6] The checklist for reptiles dwelling on Santa Cruz lists two iguanas, one gecko, and three snakes, one of which is a rattlesnake unidentified as to species. Total reptilian diversity for the island amounts to only those six, compared to twenty species that inhabit the largest gulf island (Isla Tiburón) and only two on one of the smallest (La Rasa). Island size, but also island proximity to mainland populations, are the principal constraints limiting island colonization by terrestrial species. Saltwater constitutes a significant barrier to the migration of land-dwelling reptiles, but vegetation washed out to sea during storms as floating rafts provides another vector for immigration when live cargo makes a fortuitous landfall on a previously unoccupied island.

A published extraordinary account describes how fifteen iguanas landed on the Caribbean island of Anguilla aboard a log raft on October 4, 1995, about a month after Hurricane Luis tracked through the Lesser Antilles as a Category 4 storm.[7] The green iguana (*Iguana iguana*) had never been reported on Anguilla prior to their observed disembarkation. The storm is credited both as the disturbance that uprooted vegetation on one island and for the formation of a log raft that carried its passengers some 125 miles (~200 km) to the next. Major storms were likely to have played a similar role in the spread of lizard populations to islands throughout the Sea of Cortez.

Isla San Diego lies due south of Isla Santa Cruz. It is a yet smaller granite island, amounting to merely a quarter square mile (0.6 km²) in area. Another fault block, the island exhibits a coastal outline similar to Santa Cruz with an elongated outline having steeper slopes on the east side than on the west. The maximum height of San Diego tops 525 ft (160 m). At low tide, narrow gravel beaches hug the coast below high sea cliffs. Viewed at an oblique angle from the south against a calm sea when sky and water merge as a blue backdrop, the pink island poses an image at once forlorn but strangely inviting (plate 12). To my knowledge, no one has bothered to conduct a survey of intertidal marine invertebrates around its rocky shores, but Pleistocene fossils including pecten shells and coral colonies (belonging to *Pavona gigantea*) are found cemented firmly in place among granite boulders on the island's south end. Herpetologists have canvassed the island to record a fauna of three reptiles, including the same two species of iguana found on Santa Cruz, but a different species of gecko, and no snakes. Vegetation is scrappy at best.

These two islands evoke an irresistible fascination for the geologist interested in reaching a better understanding of their relationship to one another and to the surrounding seafloor extending west to the peninsular mainland and east beneath ever-deeper waters. The key to unlocking this puzzle rests with bathymetry portrayed as a kind of topographic map charting the seafloor. An early and highly successful attempt to collect and collate data on variations in water depth throughout the Gulf of California was undertaken during an expedition sponsored by the Geological Society of America and the Scripps Institute of Oceanography from October to December 1940, outfitted on the research schooner *E. W. Scripps* and logging 4,600 miles (7,400 km) within the gulf region. The marine geologist in charge of bathymetry was Francis P. Shepard (1897–1985). Publication of the detailed bathymetry report together with the rest of an extensive memoir on the geology and paleontology of the gulf region was delayed for a decade.

When finally published (Shepard 1950), the survey included a package of detailed charts for different regions between Isla Ángel de la Guarda in the north and the southern tip of the peninsula. Chart 7 covers the area around Santa Cruz and San Diego islands. Most impressively, Shepard's report features a folded bathymetric chart for the entire Gulf of California printed on a single piece of paper 1.5 by 4 ft (0.46 by 1.22 m). The composite map identified for the first time a series of extremely deep basins, among them the Farallon basin east of Santa Cruz Island. Based on the level of accuracy for that time, it was found to be the third deepest part of the gulf at a depth of 1,750 fathoms (3,200 m). Several decades later, work by Lyle and Ness (1991) added a level of sophistication commensurate with the complexities of plate tectonics, but firmly based on the accuracy of Shepard's earlier contribution.

As modified (figure 8.5), part of chart 7 shows variations in bathymetry at intervals of 100 fathoms (~183 m) based on 19 ship tracks that crisscrossed the area with over 750 individual depth soundings. In his written report, Shepard (1950) included a photograph framing the west side of Isla Santa Cruz as a fault scarp. Hence, the chart adds to island topography (map 8) by denoting a pair of parallel faults in close alignment on opposite sides. Taken together, Santa Cruz and San Diego are at center stage on the chart, but a third feature appears as a hole in the seafloor a little farther north. The hole is not trivial, because it is similar in size and shape to Santa Cruz and marks a depression exceeding 500 fathoms (915 m) in an otherwise flat plain defined by the 400-fathom (732-m) isobath. At first look, it appears that the hole lines up on a north-to-south axis with the two islands. The well-defined depression fits the expected profile of a graben, whereas the two islands are fault-block horsts. Lacking insight from the bathymetry, the passage between the two islands might appear as a simple gap in a single north-south ridge. But the relationship is more complicated. On closer examination of submarine contours, the islands belong to separate ridges dissected by a submarine fault. The three features may have been aligned more side-by-side, but subsequently offset through regional tectonics.

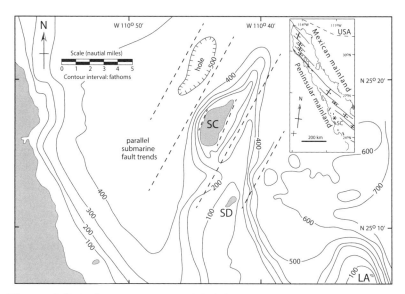

Figure 8.5. Bathymetric map modified from Shepard (1950) for the area around Isla Santa Cruz (SA) and Isla San Diego (SD) and extending southeast to include the rock at Las Animas (LA). Map inset depicts the deep-water basins at the center of the gulf with seafloor spreading zones and associated strike-slip faults.

An aspect of Shepard's insight as a marine geologist was his rhombic solution, incorporating two phases of tectonic activity.[8] Starting with a regular square having a single pair of faults parallel to one another on two sides, a graben is accounted for by downward slippage between the two faults when ocean crust is stretched apart. An adjacent setup with parallel faults may end as an island horst above the waterline. In the subsequent phase of tectonic activity, renewed movement along the same faults is no longer up and down, but side to side in a lateral motion called a strike-slip fault. The lateral slippage deforms the original box (or crustal block) into a rhomb. In effect, the configuration of the seabed between Santa Cruz and San Diego lies at a considerable water depth of 150 fathoms (275 m) and traces a twist in bathymetry related to the tectonic push against a square making it into a rhomb. The map inset (figure 8.5) shows the succession of deep basins running through the center of the Gulf of California, each with spreading centers offset by transform faults such that the entire Baja peninsula is slowly shifting to the northwest relative to the rest of the North America. The fault at the northern end of the gulf connects with the famous San Andreas Fault slicing through Alta California on the American side of the international border. It is in this context that Lyle and Ness (1991) evoke the "common class of rifts" quoted at the start of this chapter.

Even so, the orientation of submarine faults around Islas Santa Cruz and San Diego runs askew of the patterns imprinted to the north at Puerto Escondido and Isla Danzante (see chapters 6 and 7). Here, the local story is different and an additional piece of the puzzle comes from the submarine contours that sweep around the barren chip of granite called Los Animas (figure 8.5, LA, lower-right corner of the map). The bathymetric indentation between the peninsular mainland and the separate ridges on which Santa Cruz and San Diego sit is mirrored by a similar but deeper indentation to the east that sets off Las Animas. Changes in bathymetry eastward across the 600- and 700-fathom (1,100- and 1,280-m) lines suggest a tectonic shove to the northwest, perhaps in concert with the ocean-spreading center in the deep Farallon basin. At sea among these granite rocks, the whole story resonates with wrinkles yet to be appreciated.

Confession

During my epic sea journey in 2003, we planned to stop off at Islas San Diego and Santa Cruz on the way back north to Loreto from Isla San José. Our trip south began with multiple large containers of gasoline to keep the powerful twin engines at the back well supplied. Leaving from San José, we were confronted by one of those fierce January winds blowing out of the north, and fuel consumption surged in our fight against contrary ocean swells. Aiming closer to the peninsular shores, we still were forced to cross open water a hundred fathoms in depth. Our boatman informed us that if we went to Isla Santa Cruz, we might not have sufficient fuel to return all the way to Loreto. I never made it to either island, although I look for them whenever possible on flights from La Paz to Los Angeles. A great plume of gravel and boulders is reflected beneath the green tone of shallower water off the southern tip of Isla San Diego. It must be influenced by longshore currents that sweep along the flanks of that island, but also enhanced by episodic hurricanes that spiral northward into the Gulf of California. I remain most impressed by the striking east-west asymmetry of both islands that suggests to me how excessive material has been eroded off fault scarps from the east sides and dumped thereafter into deeper waters. Only the truly big storms are capable of that work.

I am fortunate in the help freely provided by informants who maintain a lively correspondence regarding my study of the gulf coast of Baja California and its islands. Images in this chapter were provided by one such informant, who was willing to visit the islands at my behest. Other field photos sent to me confirm irrefutable evidence of Pleistocene fossils from the conglomerate boulder beds on Isla San Diego. The marine invertebrates that reached these far islands in the past colonized very narrow shelves. Together with available topographic and bathymetric maps for the region, such images are what satisfy my longing for Islas Santa Cruz and San Diego.

9

Cabo Pulmo and the History of Fossil Reefs in the Lower Gulf

Clinging to the coral, growing on it,
burrowing into it, was a teeming fauna.
Every piece of the soft material broken off skittered
and pulsed with life—little crabs and worms and snails.

John Steinbeck, The Log from the Sea of Cortez (1951)

THE TROPIC OF CANCER occupies a line of latitude 23°26' north of the equator. Its counterpart is the Tropic of Capricorn, located the same distance south of the equator. During the inverse seasons in the northern and southern hemispheres, these invisible geographic circles inscribe the sun's maximum noontime position during the summer solstice before the tilt of the earth's axis causes an apparent shift in the sun's daily march across the sky. As such, Cancer and Capricorn are closely knitted to our planet's orbital relationship with the sun at the center of our solar system. In addition, the latitudes mark the limits of the tropics where sunshine is most intense and where life on land and adjacent coastal waters is biologically most diverse. The spot where the Tropic of Cancer crosses Mexico Highway 1 is marked by a roadside monument and tourist plaza. At that place, the gulf shore is 18.5 miles (30 km) due east and intersects with the village of Cabo Pulmo and Cabo Pulmo National Marine Park (figure 1.7).

Encompassing 27.5 square miles (71 km²), the marine park was launched in 1995 as a certified Natural Protected Area conserving the only coextensive coral-reef ecosystem in the Gulf of California. Moreover, the Cabo Pulmo reef is the most northern system of its kind on the west coast of North America. Corals constitute the key umbrella species that sustain a complex web of life dependent on the reef's physical infrastructure for its welfare. Reef builders are dominated by three coral species (*Pocillopora elegans, Porites panamensis,* and *Pavona gigantea*) that form a stony-interlocking framework as much as 6.5 ft (2 m) in relief, within which a multitude of other marine species find shelter. Some, including specialized sponges (*Cliona* sp.), molluscan bivalves (*Lithophaga* sp.) and barnacles (*Acanthemblemaria crockeri*) bore into the corals and live mainly concealed within, whereas most other marine invertebrates dwell outside but within the reef's cavities. An early survey of the Cabo Pulmo reef system by Brusca and Thomson (1975) lists five resident sponge species, eight corals, two sea fans, two hydrozoans, seven marine worms, twenty-four gastropods, three bivalves, five barnacles, twenty crabs, five starfish, five ophiuroids, five sea urchins, and nine sea cucumbers. It is estimated that 26 percent of the 875 known fish species in the Gulf of California dwell in and around the Cabo Pulmo reef. A more current overview places an emphasis on the reef builders and their biology and summarizes what is known about corals elsewhere in nonreef settings throughout the gulf (Reyes-Bonilla and López-Pérez 2009).

This chapter describes what can be seen onshore during an easy ramble across the beach and rocky coast adjacent to the reef. Commercial guides under park supervision operate from the center of the village, and a visit to the main reef for scuba diving or snorkeling is highly recommended. As a geologist, however, my attraction to Cabo Pulmo is predicated on what can be learned about the Pleistocene history of the reef and how it came to thrive precisely at this place. The excursion is at a relaxed pace across mostly even ground and bears repeating to best effect at low tide during a few days' visit. Fossil coral reefs occur far to the north, and their backstory provides insight on the potential near-term effects of global warming in this part of the world in the context of global Pleistocene temperatures that were higher than today.

Hiking Ashore the Coral Reef at Cabo Pulmo

During the 1950s, a major study of the Gulf of California's marine biology was mounted as the "Puritan-American Museum of Natural History Expedition to Western Mexico." Extensive collections of marine invertebrates were made during a 40-day cruise by the research vessel *Puritan* and the results published through a series of articles in the museum's official bulletin. A summary of living and fossil corals throughout the gulf region appeared as the seventh article in that series (Squires 1959). The coral reef at Cabo Pulmo became the central focus of the report, in which it is claimed that the only other scientific description was made by Steinbeck and Ricketts (1941) in the account of their private excursion aboard the *Western Flyer* in 1940. Indeed, the pair stopped at Cabo Pulmo for a single day on March 18, 1940, and regretted not having planned ahead to include snorkeling equipment for the trip.[1] With the *Western Flyer* safely anchored well off the reef, Steinbeck and Ricketts went ashore in the ship's skiff. At low water, tidal pools could be explored in the back-reef zone, and a bounteous collection of marine invertebrates was sampled. As an experiment, a few corals were taken back to the ship and allowed to sit in a pan of seawater that slowly grew stale from lack of oxygen. Observed during this process, "thousands of little roomers" that live within the interstices of the coral came out of hiding and were easily picked off for closer study.[2]

During the *Puritan*'s visit, time seemed to have stood still or perhaps even regressed in terms of local commerce. Squires (1959) interviewed a former pearl diver, Jesus Castro Fiol, who had lived at Punta Cabo for 35 years and was able to share much about the area's once thriving pearl-fishing industry. Based on the informant, members of the *Puritan* expedition produced a map over the 6 square mile (15 km^2) bay between Cabo Pulmo and Las Frailes, showing a series of named reef tracks that follow rocky bars parallel to the coast. Because reef growth traces the submarine topography of the rocks at regular intervals, the tracks are not regarded as part of a true fringing-reef system. Even so, reef growth is luxuriant, and the technical distinction seems, at best, to be inconsequential. Steinbeck, Ricketts, and

Squires arrived at Cabo Pulmo by sea, when nothing much of human activity was evident ashore. Today, a thriving community has grown in place that benefits directly from ecotourism related to the Cabo Pulmo National Marine Park. The village may be reached by the coastal gravel road from the tip of the Baja California peninsula, but access is more efficient by way of paved Mexico Highway 1 and its byway at the town of La Ribera located 15.5 miles (25 km) north of Cabo Pulmo village (map 9).

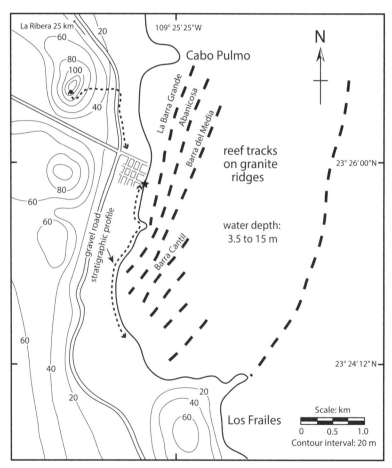

Map 9. Topography around the village at Cabo Pulmo and layout of contemporary coral reefs aligned with offshore rocky ridges in the Cabo Pulmo National Park. Star marks launch area for dive tours.

Timing is everything at Cabo Pulmo, and the day must be planned with attention to the tides. What cannot be achieved to best effect on any given day may be put off to the next. The two events featured herein are divided between a half day at the shore and a half day hiking to the summit of the 475-ft (145-m) promontory northwest of town. There is no urgency for a rushed visit to Cabo Pulmo, as the inherent easiness of the place is conducive to a stay of several days, especially if scuba diving or snorkeling is added to the agenda.

From the national park's information kiosk at the shore in town, a magnificent white-sand beach stretches south unobstructed for a third of a mile (0.5 km). Irrespective of the tide, the beach offers a peaceful esplanade that may be enjoyed any time of day, although the early morning is especially attractive, with the sun's illumination from the east. During their March visit, the shallow water behind La Barra Grande close to shore was so warm that Steinbeck and Ricketts pulled off their wading boots and donned tennis shoes to protect their feet. Like the beach, the seabed is covered by carbonate sand derived from the slow disintegration of corals and other shelly marine life. Purple-and-gold sea stars (likely *Phataria unifascialis*) were noted as abundant, and the back-reef zone was littered with knobby corals in waist-deep water.

Continuing southward with a receding tide, the observant beach walker will encounter patches of conglomerate that rise above the exposed sand-like tiny islands. The conglomerate is cemented and consists of well-rounded cobbles and pebbles eroded from andesite and granite bedrock. Waves pushing westward across the bay turn to white foam where they break over ribs of conglomerate oriented parallel to the shore. The beach makes a slight turn more to the southwest on the other side of a prominent spur jutting into the bay. After another third of a mile (0.5 km), the sand beach ends, and mounds of pink granite form a rumpled canvas that expands ever larger in size with the falling tide (figure 9.1). The parts that usually remain dry above normal high tide are rosy pink from the mineral plagioclase, offset by flecks of black biotite and neutral grains of clear silica. Lower a little farther offshore, the wet granite casts a darker complexion. I am able to spend hours here, although limited by the tidal cycle. The living reef ecosystem has its own attraction

Figure 9.1. Shore exposure of Cretaceous granite surrounded by Pleistocene conglomerate composed of granite and andesite cobbles. Photo by author.

and I could be happy to strap on a pair of goggles and wade chest deep into the water. But my affinity is for fossil hunting, and it casts a strong spell that pulls me back through prehistoric time to the late Pleistocene.

Subtle to the eye at first, there exists a former ecosystem hereabout that makes its presence known through fossil remains. It requires patience to visualize, but the life detailed from a time more than 100,000 years ago gradually comes into focus. The first clue is found along the side of a granite mound, where a line of white-shelled oysters (*Ostrea chiliensis*) is attached to the rock in growth position (figure 9.2). The two valves that make a complete oyster shell are not a perfect match. There is a small difference in shape between the two, and it is the somewhat flatter right valve that is fastened to the

Figure 9.2. Pleistocene oysters (*Ostrea chiliensis*) attached in growth position to a granite surface (coin is 1 in, or 2.5 cm, in diameter). Photo by author.

rock. Here, we see only individual, right valves stuck to the granite, because the opposite valve separated and washed away when the ligaments holding the two valves together deteriorated after death. The fossil oysters are large, 6 in (15 cm) lengthwise and aligned uniformly side-by-side against the granite surface. One can imagine many more oysters were originally attached to the rocks, but gradually were lost due to the abrasion of wet sand.

Vertical joints in the granite are the loci for ongoing erosion. Some are enlarged as deep furrows in the rock, smoothed with time to leave the visual impression of a row of bread loaves sitting side by side on a cooling rack. Other joints oriented more perpendicular to the shore absorb a pounding by the waves, particularly during storms. These have a tendency to widen into narrow alleys that become traps not only for sand and pebbles but also for biological detritus. Such erosional crevices provide yet another fruitful context for exploration by the fossil hunter. One of the bigger crevices in the

Cabo Pulmo granite is a repository for large colonies of branching coral (*Porites panamensis*) that are wedged into place (figure 9.3). The corals did not grow here, because they are clearly caught in a sideways position. What must be assumed is that colonies were torn from a Pleistocene reef during a storm and slammed ashore with sufficient force to become tightly wedged in place.

The conglomerate cover above the granite also is a rich repository of Pleistocene fossils. Oysters commonly occur in growth position seated atop the conglomerate surface a little below those noticed earlier attached to the granite (figure 9.2). At the time of oyster colonization, individual clasts within the conglomerate were cemented

Figure 9.3. Large Pleistocene coral colony (*Porites panamensis*) wedged in a crevice eroded in Cretaceous granite. Photo by author.

together into solid rock. The oyster made no distinction between granite as an attractively hard surface and what already was a solid pavement of conglomerate. The difference between the top of the granite and the top of the conglomerate amounts to little more than 18 in (~45 cm), which implies that, at least locally, the water depth remained quite shallow (figure 9.1). On closer inspection, however, other fossils are interred within the conglomerate and must have arrived before all the clasts were cemented together as a solid mass.

A wider survey of the conglomerate layer is thus necessary in order to get a feeling for an earlier phase of development prior to oyster colonization. With a systematic search at low tide, several bits of data may be added to the story. Especially for coral colonies with a sub-spherical growth form, it can be predicted that a storm easily might pluck them from the reef and launch them landward. Indeed, just this sort of coral head (*Pavona gigantea*) can be found as a biological clast buried in the conglomerate. Some are relatively large, up to 12.5 in (~32 cm) in diameter. Likewise, fragments of the branching coral (*Pocillopora elegans*) occur within the conglomerate. The softer calcium carbonate composition of corals is more prone to breakage and erosion than the harder granite or andesite clasts in the same deposit. Indeed, the fossil corals within the conglomerate appear weathered.

A diligent search locates additional examples of fossil remains subjected to a battering at the time of their burial. Large marine gastropods (*Strombus galeatus*) commonly occur within the conglomerate, but their shells are broken (figure 9.4). Under the tumult of igneous clasts jostled one against the other by storm waves, it is not hard to understand how the sturdy shell of a stromboid sea snail might be degraded. Not as obvious, the large shells of spiny oysters (*Spondylus princeps*) also occur within the conglomerate, but are typically visible only in cross section as curved lines. All these fossils share extant descendants in the Gulf of California that confirm an intertidal to very shallow-water habitat.[3] Hence, the Pleistocene ancestors probably were not far removed from their preferred setting after postmortem transport in a landward direction. The overall impression from our survey is that a first phase of development brought together an influx of cobble-size granite and andesite cobbles mixed with organic remains as a shallow deposit that failed to fully cover

Figure 9.4. Eroded marine gastropod (*Strombus galeatus*) revealing the central spire in cross section (coin is 1 in, or 2.5 cm, in diameter). Photo by author.

granite basement rocks. Once the conglomerate was stabilized by cementation, the second phase of development took place with the arrival and colonization by robust oysters (*Ostrea chilensis*). It is left to the imagination, but an extensive oyster bank may have rapidly populated the rocky shore during a pause in storm activity lasting only a few years.

The last vestige of beach sand deposited between rocky highs in granite and associated conglomerate is met a half mile (800 m) beyond the turn in the shore, where the prominent spur points seaward (map 9). Where the granite rises well above sea level and the basal conglomerate is at its thickest offers a superior spot to measure a stratigraphic column showing relationships between successive rock layers (figure 9.5). The junction between granite and overlying

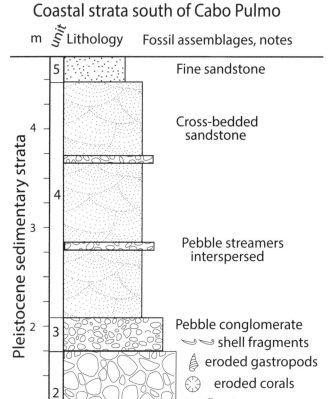

Figure 9.5. Stratigraphic column showing the succession of igneous and sedimentary rocks representative of the coast south of Cabo Pulmo. Section location is marked on map 9.

conglomerate denotes an unconformity with a highly irregular surface. In effect, the conglomerate was deposited in a step-wise fashion during a rise in relative sea level that encroached against a rocky shore.

Here (figure 9.5, unit 1), the maximum extent of granite visible above the water level amounts to 4 ft (1.25 m). The overlying coarse conglomerate that includes eroded corals, spiny oysters, and marine gastropods is variable in thickness (unit 2), but at best attains an equal thickness. Above the coarse conglomerate, there appears a sharp transition to a thin pebble conglomerate about 1 ft (30 cm) in thickness (unit 3). The layer includes a fossil fauna dominated by the shells of mollusk bivalves and a rare gastropod. Those most readily identified as to species include the bittersweet (*Glycymeris maculata*), venus (*Chione californiensis*), and cockle (*Cardium biangulatum*) shells. A marine snail with a typically oval shape is recognized only to genus level as an olive shell (*Oliva* sp.).

Next in order (figure 9.5, unit 4), the pebble conglomerate is followed by approximately 8 ft (2.5 m) of cross-bedded sandstone interspersed with pebble streamers. Low angles in the cross-bedded sandstone are indicative of shifting sand on a Pleistocene beach interrupted by occasional events that introduced pebbles and finely crushed shell material from offshore. The entire sequence is capped by 1 ft (30 cm) of fine sandstone lacking any trace of cross-bedding (unit 5). Superimposed on Cretaceous granite, the principal signature of the Pleistocene rock layers indicates a back-reef setting pinned between a coral reef in the seaward direction and a rocky coastline. Ultimately, the granite shore was replaced by a stable beach that suffered only minor disturbances. The rock column (figure 9.5) records events that transpired at this place over a few thousand years, beginning roughly 100,000 years ago. A short distance to the south, pink granite in wonderfully sculpted shapes is exhumed from beneath the cover of Pleistocene sandstone found juxtaposed in cliffs that rise above in the background (plate 13). Nature, it seems, is capable of competing with the human hand in the creation of arresting art.

The recent past and present represent closely matched environments at Cabo Pulmo. As time permits and to gain a closer perspective on contemporary rocky-shore dynamics, we may continue onward to the south for another mile (1.6 km). Along the curve in the shore toward the headlands of Los Frailes (map 9), the granite coast rises in relief and the uneven surface is more challenging. I am content to stop at a suitable resting place where the slap of the waves against the rocks may be felt at close quarters. During low tide, sea snails with coarse ribs that trace low spires (*Nerita scabricosta*) occur in heaped clusters well above the waterline (plate 14). I have found their distinctive fossil ancestors on a Pliocene rocky shore raised high by gulf tectonics farther to the north at Isla Monserrat,[4] and I thrill to see their descendants alive here in prolific populations. These marine invertebrates from the upper intertidal zone are accustomed to remain exposed to the air for hours at a time, and they are among the largest of their kind found anywhere in the Gulf of California.[5] It will take some hours for the sea to return to this level.

The trek back to the village center is less than a mile and a half (2.25 km), leaving ample time in the day for a shorter but more strenuous hike to the top of the prominence on the other side of the village (map 9). Like Cabo Pulmo itself, the hill is composed of rust-red rocks that align with Miocene andesite. The crest is accessible via a wide-cut path and affords a sweeping view out over the Cabo Pumo embayment, including the modern beach and rocky shore in the distance toward Los Frailes. When a steady breeze blows onshore out of the southeast, waves crossing the multiple bars on which the living reef resides mark their locations in white foam.

APPRAISAL OF LATE PLEISTOCENE CORAL-REEF EXPANSION

In addition to a Pleistocene fauna that foreshadowed the living reef at Cabo Pulmo, there exist as many as eight locations around the Gulf of California farther to the north where fossil reefs are preserved in exquisite detail (figure 9.6). Due to early survey work (Squires 1959), many but not all of these fossil localities were recorded with the proposal that global temperature was significantly warmer than today during the last interglacial epoch 125,000 years ago. Moreover, global sea level stood as much as 20 ft (6 m) higher at that time (review chapter 6). Surplus water was fed into the world's oceans from the melting of vast continental glaciers in North America and Europe, but the next drawdown in sea level reached a maximum 33,000 years ago during the last glacial epoch when global sea level was more than 328 ft (100 m) lower than

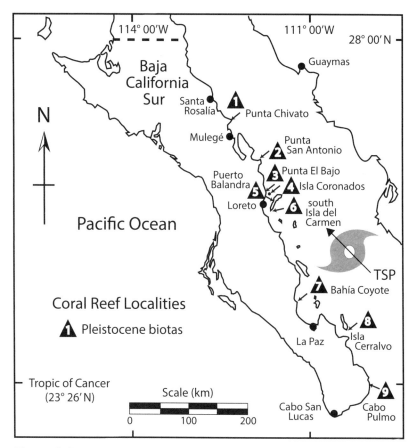

Figure 9.6. Map of Baja California Sur showing coral-reef localities that preserve Pleistocene biotas. Pin-wheel symbol labeled TSP denotes a theoretical Pleistocene tropical storm path.

today.[6] It must be emphasized that sea ice never impacted the Gulf of California during the last or any of the earlier Pleistocene glaciations, although the region received more rainfall during glacial maxima than today. Indeed, over the last 800,000 years there have been many such cycles in climate that resulted in major perturbations in global temperature and sea level. The upper Pleistocene fossil reefs found around Gulf of California are special, because they remind us of Planet Earth's last major phase of global warming. It also may be argued that the former reefs were mostly limited to sites sheltered from Pleistocene hurricanes.

All such Pleistocene reefs are confined south of the boundary between the northern state of Baja California and Baja California Sur, demarcated at 28° north latitude. The most northerly example occurs in an alcove on the southern side of the Punta Chivato promontory 285 miles (460 km) northwest of Cabo Pulmo (figure 9.6, locality 1). Punta Chivato is slightly more than 3.5° of latitude farther north than Punta Pulmo, which on a strictly north-south axis amounts to a distance of about 242 miles (390 km). Based on detailed vertical and longitudinal mapping through the center of the Punta Chivato reef over a distance of 230 ft (70 m), three generations of reef growth by *Porites panamensis* are known to have supported an ecosystem of 50 marine invertebrate species.[7] Although relatively small in area, the preserved biodiversity of the Punta Chivato reef exceeds that of any other fossil reef in the Gulf of California. Individual coral colonies subsist today offshore Punta Chivato, but they fail to coalesce as a ridged structure capable of attracting a dependent ecosystem. Our assumption is that the average year-round water temperature at Punta Chivato must have been comparable to that of Cabo Pulmo today, 125,000 years ago in order to support a thriving reef ecosystem.

Farther south near Punta San Antonio (figure 9.6, locality 2), a small Pleistocene reef with a single generation of *Porites panamensis* colonies is sheltered at the end of a narrow, east-facing cove (Johnson and Ledesma-Vázquez 1999). Little wave agitation reached this geographic cul-de-sac, where the coral colonies adapted a more massive morphology. The biodiversity of associated marine invertebrates stands at eighteen species. A north-facing rocky coast of Punta El Bajo 5 miles (9 km) north of Loreto (figure 9.6, locality 3) is the site of a linear reef structure stretching for 165 ft (50 m), also consisting of a single generation of *Porites panamensis* colonies, some of which adapted heavy, club-shaped branches likely in response to strong wave action. The maximum preserved biodiversity of coinhabitants is found to be only 8 species of marine invertebrates.

The most extensive Pleistocene reef in the lower Gulf of California covers an area of 3.10 acres (1.3 ha) on the south side of Isla Coronados (figure 9.6, locality 4). The reef developed in a lagoon sheltered behind a bar of Miocene andesite with a single, small tidal entrance and consists of a single generation of *Porites panamensis*, some colonies of which achieved a remarkable height of 43 in (1.1 m).[8] Although as many as five species of marine invertebrates are associated with the initial colonization of the reef structure on a cobble-strewn surface, a pattern of tight branch growth among the colonies appears to have limited sanctuary space to three species of mollusk inhabitants.

As yet poorly studied, two sites are known from Isla del Carmen (figure 9.6, localities 5 and 6) where Pleistocene reef structures remain intact. The more substantial buildup is at Puerto Balandra, a small west-facing bay on the island's northwest coast. A narrow opening to the Carmen Passage makes the embayment well protected from strong wave action. Multiple generations of short-lived *Porites panamensis* colonies amount to 20 ft (~6 m) of reef limestone. Preserved in growth position, coral density is high, with little or no trace of a related marine invertebrate fauna. A much smaller reef structure consisting of two generations of *Porites panamensis* colonies occurs at the back of a narrow inlet on the south side of Isla del Carmen, where the only evidence of an associated marine fauna consists of worm tubes up to 6 inches (15 cm) in length cemented to coral branches.

Multiple Pleistocene coral mounds extend intermittently for a distance of 9.3 miles (15 km) along the shores of Bahía Coyote north of La Paz (figure 9.6, locality 7). The reefs were formed by a combination of *Porites* and *Pocillopora* corals on a foundation of coarse sand and oyster debris (DeDiego-Forbis et al. 2004). Recurrent generations of corals with a maximum colony height of 20 in (~50 cm) accreted one on top of the other to a height of 26 ft (8 m). The diversity of marine invertebrates found to coinhabit the reefs was limited to only a few mollusk species. Related deposits interspersed among the reefs and isolated oyster mounds include abundant rhodoliths (concretions of coralline red algae) and small *Porites* colonies mixed together with a more diverse molluscan fauna of gastropod and bivalve species.

Closer to Cabo Pulmo, the southwest end of Isla Cerralvo features five recurrent cycles of reef growth established by a combination of *Porites* and *Pocillopora* corals (figure 9.6, locality 8). The paleoecology of this site was surveyed on an exposed planar surface covering

36 square yards (30 m²), where a comprehensive population census of coinhabitants was possible (Tierney and Johnson 2012). For example, the ratio of *Porites* to *Pocillopora* corals preserved in growth position was found to be nearly 10:1, but the diversity of mollusk species sheltered within the framework was dominated by no more than four of the larger gastropods and bivalves. Each successive cycle was extinguished by an event involving total burial by igneous cobbles carried from the island's interior during major storms.

The potential for a major tropical storm like Hurricane Odile in 2014 to destroy Pleistocene reefs in the lower Gulf of California is predictable on the basis of geographic location and orientation. Reef structures at Punta Chivato, Isla Coronados, Puerto Balandra, and Isla Cerralvo (figure 9.6, localities 1, 4, 5, and 8) occur in spots that were especially well protected from wave surge arriving from the east stimulated by a passing hurricane. The structures most vulnerable to wave surge occur on the north-facing shores of Punta El Bajo and east-facing shores of Bahía Coyote (figure 9.6, localities 3 and 7). Smaller reef structures near Punta San Antonio and the south side of Isla del Carmen (figure 9.6, localities 2 and 6) were somewhat less precarious due to narrow inlets where a direct hit would be required to cause damage.

Be that as it may, most of the Pleistocene reef structures in the lower Gulf of California underwent repeated burial and regeneration in the same places. Although storm surge failed to obliterate any such reefs, the heavy rain that accompanied major storms loosened and mobilized a torrent of coarse gravel inland that buried them in their original growth position. If not buried by outwash gravel, as in the case at Puerto Balandra, it may be that the influx of fresh water ponded over a span of days was responsible for repeated coral die-offs. It is notable that none of the Pleistocene examples occur on the east side of the many islands in the Gulf of California, where wave surge from passing hurricanes is predictably most intense.

Fluctuations in global temperature and changing sea level that accompanied the advance and retreat of continental glaciers in the northern hemisphere over the last 800,000 years occurred gradually, with the longest spans between the recurrence of glacial and interglacial epochs operable on a 100,000-year cycle. We are now engaged in a profound environmental experiment, during which the average global temperature shows signs of elevation at a rate far exceeding anything known to have occurred under normal conditions regulated by our planet's orbital relationship to the sun. It is not unreasonable to ask how quickly living coral reefs might return to Punta Chivato as the pattern of global warming continues unabated.

10

Pacific Bound and Coming to Terms with the Future

That night it rained. But this was no ordinary shower
or even a downpour, this was an El Niño event, evil
harbinger of the apocalyptic weather to come.

T. C. Boyle, A Friend of the Earth (2000)

Dystopian descriptions of society projected some time into the future as a warning of possible deprivations to come belong to a genre all its own in literature and in film. Most focus on drastic political change, whereby a totalitarian regime exerts control over populations no longer capable of self-determination. Hegemony in many such fictional plots takes advantage of weaknesses in the human condition that are internal in character. Quite another prospect is the dystopia that stems from external factors imposed by global climate change. With *A Friend of the Earth* (Boyle 2000), life portrayed in southern California in the near future sketches a dark but comic portrait of good intentions by environmental activists to influence society before it becomes too late. At what point are societal bonds dissolved when the services we come to expect in our daily life fall into disrepair? At the most rudimentary level, when does our network of roads and bridges become so wreaked by the forces of nature that recovery is daunting on more than a regional scale? A state sufficiently rich and large in geography allocates resources so that the calamities of one region may be alleviated with outside support from other parts.

The Frontier States of Baja California are the least populated in all of Mexico. Construction methods have improved over time. State and federal support for the rebuilding of roads and bridges has been adequate to keep the fabric of society smoothly running for a relatively small population at the mercy of rare but devastating weather events. Not all governments are equally endowed with assets for renewal after debilitating natural disasters. Examples from the near and more distant past in Central America are informative. Seemingly unprecedented, the intense rainfall from Hurricane Mitch in 1998 resulted in widespread landslides throughout Honduras that degraded 70 percent of that country's roads, caused 7,000 human fatalities, and left more than 3 million homeless.[1] The storm made landfall on October 29 from the Caribbean side of the country as a Category 5 hurricane that stalled for more than a week, dumping as much as 50 in (1.25 m) of rain across the interior. Archaeological evidence from a more distant past 900 years ago suggests that the pre-Columbian Mayan culture in present-day Honduras also was vulnerable to storms that triggered mudslides and debris flows capable of burying settlements (E.M. Johnson et al. 2020). Where the resources of a state are limited, there is little possibility of recovery from extensive natural calamities without external aid. When does the repetition of the rare 100-year or 1,000-year natural disaster become normalized as annual events? Dire predictions posit that the unmitigated effects of global warming threaten to make Planet Earth increasingly uninhabitable for humans by the end of the twenty-first century (Wallace-Wells 2019).

Our focus has remained on the geologic record of hurricane-strength storms over the Gulf of California dating from the Pliocene to the relatively recent past of the Pleistocene 125,000 years ago to the Holocene barely 10,000 years ago. The obligation of this concluding chapter is to shift the narrative to the opposite shores of the Baja California peninsula facing the open Pacific Ocean. A growing number of historic hurricanes appear to have entered the Gulf of California during the last 7 decades (review figure 1.5), all emanating from a zone in the eastern Pacific Ocean off the coast of Acapulco around 15° north latitude before blowing into the enormous cul-de-sac represented by the gulf.

The Baja California peninsula itself is a feature welded to the tectonic Pacific Plate but still linked to tectonic activity alongside the adjacent North American Plate through the famous San Andreas Fault. As the gulf's very origins are linked directly to the geologic history of the Pacific Plate, it is only natural to consider the powerful hydrologic cycles also in operation over the vast Pacific Ocean basin. For its permanent residents, there may be a sense of isolation living on a peninsula that stretches 745 miles (1,200 km) from Alta California in the northwest. With the potential normalization of El Niño events analogous to the Pliocene Warm Period of some 4 million years ago (Warra et al. 2005; Brierley et al. 2009), it may be that residents throughout all parts of the Baja and Alta Californias share much in common with those who dwell in even larger population centers on the far side of the Pacific rim.

FEATURE EVENT

Crossing to Todos Santos on Pacific Ocean Shores

Land's End at Cabo San Lucas, with its iconic sea arch etched in granite, is but a minuscule toenail appended to the stout limb of the Sierra de la Laguna, rising to an elevation of 5,111 ft (1,558 m) at Picacho San Lazaro little more than 20 miles (32 km) north of the peninsula's tip. At the center of the protected Sierra de la Laguna Natural Park some 40 miles (64 km) from the cape, the mountains reach a maximum elevation at 7,096 ft (2,163 m) above sea level. Departing from Mexico Highway 1 a short distance north of Los Cabos International Airport (map 10), a secondary road quickly turns into a gravel road crossing into the mountains north of Picacho San Lazaro. The distance along this route from one side to the other amounts to little more than 25 miles (40 km), but follows a tenuous track seldom attempted by the throngs of ordinary tourists who land at Los Cabos International Airport. Initially, the grade upward through a long and narrow valley penetrating westward is not intimidating.

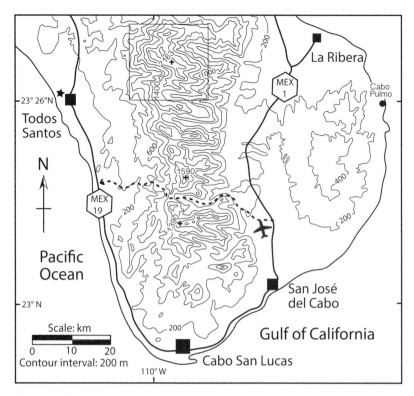

Map 10. Topography of the Baja California peninsula at its southern end, featuring the Sierra de la Laguna and high peaks between the Gulf of California and Pacific Ocean. Box defines the Sierra de la Laguna National Park. Star marks location of the Sierra School on the outskirts of Todos Santos.

Roughly 12 miles (19 km) into the sierra, it comes as a surprise to encounter a large boulder announcing in Tibetan script the entrance to Tsegyalar West. Here is located a 3,000-acre (1,215-ha) retreat dedicated to the Dzogchen tradition of Buddhism. I am instantly reminded that earlier visitors to the Gulf of California such as Ed Ricketts and Ray Cannon found solace in aspects of Buddhism.[2] The Orient is geographically far off on the other side of the Pacific Ocean, and yet philosophically nearby.

The southern pass through the Sierra de la Laguna is reached halfway across the peninsula just before a sharp bend in the road at an elevation of 2,592 ft (790 m) above sea level (map 10). The aforementioned Picacho San Lazaro is within sight to the south, and the higher peak at Picacho La Zacatosa at 5,217 ft (1,590 m) rises a short distance to the north. Rounding a bend at the side of the pass, we arrive at a view point where the deep blue of the Pacific Ocean is slotted above a widening valley descending to the west coast (plate 15). The high country hereabout is home to plants and animals different from the surrounding, coastal desert. For those attracted by the Cape Region's botany, birds, and mammals, extensive checklists are readily available (Zwinger 1983). A distinctive oak forest (*Quercus devia*) is established at elevations between 3,900 to 5,200 ft (1,200 to 1,600 m), giving way to a mixed forest of oak and endemic pinyon pine (*Pinus lagunae*) at higher elevations.

For my part, the region's geology and hydrology are of utmost interest. Compared to the coastal desert around Puertecitos and Volcán Prieto (see chapter 1) with the lowest average rainfall on the Baja California peninsula at 2 in (5 cm) per annum, the Cape Mountains at the distal end of the peninsula enjoy the highest average rainfall at 17 in (45 cm) per annum.[3] The extra moisture over this mountain district of roughly 190 square miles (~500 km²) arrives mostly in August and September and is directly related to the circulation of subtropical storm systems. In contrast to the steeply stratified andesite highlands of the Sierra de la Giganta (see chapter 5), these mountains are dominated by granite. A rock cutting at the side of the pass gives access to the black-and-white (salt-and-pepper) granite typical of the sierra. Otherwise, relatively thick forest vegetation obscures the rocks. Another 12 miles (20 km) of dirt road lies ahead on the descent to the Pacific Ocean (map 10), and the first few after the pass are steep and deeply rutted. The rocks are well exposed at the side of the road only where the seasonal rush of water has cleared vegetation around waterfalls. The first such clearing appears near San Pedro de la Soledad after a descent of about 1,280 ft (390 m) from the pass (figure 10.1).

At a distance of 6 miles (9.5 km) from the coast at a much lower elevation around 655 ft (200 m), the valley widens and the streambed meanders through low gorges where the water continues to flow even during drier months of the year. One such gap is encountered on the drainage of the San Jacinto. Visited in February 2017, less than six months after the passage of Hurricane Newton, a Category 1 event

Figure 10.1. Granite exposure in a streambed high on the west slope of the Sierra de la Laguna at an elevation of 1,312 ft (400 m) near San Pedro de la Soledad. Photo by author.

accompanied by wind speeds of 90 mph (150 km/hr) and heavy rain, the high-water mark left by local flooding was still evident. What appears as a brown line painted at a consistent level across granite walls recorded a high-water mark 4 ft (1.25 m) above normal (plate 16). The mark shows that muddy water was impounded for enough time to leave a dirty rim before the water drained away. Little structural damage was suffered in the Cape Region as a result of Hurricane Newton, but the storm tracked northward through the peninsula into the Gulf of California (figure 1.2), where significant road damage occurred at Mulegé and Santa Rosalia due to torrential rains from the stalled storm system that fell during a continuous eight-hour period.[4] Mexico Highway 1 was washed out in several places, resulting in the temporary isolation of those gulf towns.

North on Mexico Highway 19 toward the town of Todos Santos (map 10), a series of closed, narrow lagoons occupy the low ground between the paved road and sand dunes along the Pacific coast. The lagoons are drained during the dry summer months, but during the hurricane season they serve as a kind of safety valve to absorb excessive flooding from the uplands. A stop along the highway to investigate one such feature is worth the effort. It is possible to walk to the center across punky ground without sinking too deeply into mud. On close examination, the soft surface is draped by a rumpled texture with small mounds barely an inch (2.5 cm) in diameter. These lagoons host extensive microbe communities that thrive especially well on the damp ground between flood events, somewhat like mold. The setting is rather like the damp patch of ground enclosed by storm boulders at Ensenada Almeja (see chapter 4). Much remains to be done to enlarge the study of lagoon microbes in the Gulf of California to the level of research accomplished at the salt lagoon close by Volcán Prieto (see Chapter 2). Whereas the microbial deposits at the salt lagoon extend well below the surface, those at Ensenada Almeja and here on the Pacific coast appear to be more surficial. Doubtless, the rainfall gradient between north and south plays some role as yet to be determined.

Over the years since my first visit to the Cape Region, commercial enterprises have slowly crept northward from Cabo San Lucas toward Todos Santos. Based on my own travel experiences in Mexico's Yucatán and parts of Central America, what most startles me is the transplantation of a mock Mayan village to the Cape's Pacific shores in the form of beach bungalows. To be sure, the Mayan culture imported shells of the spiny rock oyster (*Spondylus calcifer*) from sources as distant as the Pacific coast of present-day El Salvador, Guatemala, and southwestern Mexico to their lowland cities closer to Caribbean shores. Maya influence through trade of any sort with the Baja California peninsula is purely fanciful.

Todos Santos is a delightfully authentic town with a population of less than 6,500 that sits on the Pacific coast just above the Tropic of Cancer nearly equivalent in latitude to Cabo Pulmo on the opposite gulf shore (see chapter 9 and map 10). Established as a church mission in 1724, the climate proved beneficial to the development of sugarcane plantations. The town's cultural center features a fine collection of photographs from the turn of the twentieth century that captures the place and its people with a keen sense of vibrancy, but also one deeply imbued in isolation. Of late, the town has attracted artists and art galleries, not to speak of fine restaurants, that offer much to tourists seeking a reprise from the commercialism of Cabo San Lucas. An invitation in 2017 to participate in the community's annual lecture series organized around the natural sciences also provided me with the opportunity to renew contact with a former student, Tom Eckman, at the Sierra School in Todos Santos. He was among the earliest students from Williams College to accompany me to Baja California during annual January excursions in the 1990s, and he now teaches science at the private, bilingual school for Mexican and international youth at the secondary level.

Part of my duties as a guest of the community was to organize a local science hike, which readily doubled as an opportunity to engage Tom's students at the Sierra School. With a career devoted entirely to college-age students behind me, but memories of what it was like to endure the maelstrom of influences that barrage a middle-school student, I wondered how I might connect with Tom's students. Teaching geology in the classroom is an intellectual torture, when what is most demanded by the discipline is to bring pupils outdoors to encounter rocks face on. In this case, all residents of Todos Santos, including the student population, enjoy the good fortune to dwell between the Pacific shore in one direction and the foothills of the Sierra de la Laguna in the other. In short, nature abounds, and it is impossible to avoid things of interest during a stroll in any direction outside the town's central grid. The answer to my dilemma was to speak clearly, to connect science with the experience of daily life, and to get outdoors as quickly as possible for a morning stroll during the cool of the day.

The Sierra School of Todos Santos occupies a comfortable, two-story building on the outskirts of town (map 10). Passing directly in front of the school building is a dirt road intersecting with a normally dry streambed that empties westward into one of the many lagoons fronted by sand dunes at the coast. In this case, the road cuts downward into the soil and rocks to intersect with an arroyo

that runs to the ocean. Descending along the road grade to reach
the streambed below, the hiker encounters a wall of conglomerate
with a mixture of granite cobbles and boulders suspended in a matrix
of sand and clay. That is to say, the well-rounded, individual granite pieces are not in direct contact with one another. The technical
term for such a deposit is a debris flow. Although the foothills of the
sierra are more than 3 miles (5 km) distant, the debris flow represents
a major washout from the mountains caused by high-impact rain
events during the last few thousand years. As all the students had
experienced a lesser storm event with Hurricane Newton only a few
months prior, the concept of running water and its power to alter
the landscape is not lost on them.

Turning the corner at the bottom of the road, the open streambed leads to the center of the lagoon within sight. Some of the
boulders stranded in the streambed are large, too big to be rolled
downstream by a moderate flow of stormwater to the center of the
lagoon. Observing the embankments on opposite sides of the arroyo,
however, it is apparent how the turbulence of floodwater loosened
granite clasts from the soft sand that encased them to tumble inward.
In other words, the effect of gravity also plays a role in the recycling
of boulders and their introduction into the streambed. Only a few
smaller boulders make it all the way to the heart of the lagoon, where
the arroyo broadens out to intersect a shallow clay pan still awash
in muddy water. Here, a line of brushwood lies festooned along the
upper sides of the arroyo. Floated into place from upstream by the
last flood, the woody debris signal just how high the floodwater rose
at this place during Hurricane Newton. With little effort, it would be
possible to calculate the volume of floodwater captured by the lagoon
during the worst of the storm.

The streambed fails to breach the sand dunes all the way to the
sea, probably because the size of the lagoon offers sufficient accom-
modation space to absorb the rush of stormwater. The late morning
sun is strong, but determined hikers skirt the mud pan to climb a
low rampart of sand in order to reach the beach at the Pacific shore.
Here, the rampart sits below the sand dunes on either side, like a
topographic saddle between the lagoon and the descending beach.
Wet sand from the last high tide shows that ocean water lapped onto

Figure 10.2. Students on a field excursion near the Sierra School in Totos Santos in February 2017. Photo by Thomas Ekman.

the saddle and nearly succeeded in washing over into the lagoon only hours earlier. Our topographic saddle is a geographic meeting point in an ongoing battle between the daily and monthly vicissitudes of the tides and the episodic arrival of floodwaters related to big storms. Under the leadership of their science teacher, students from the Sierra School of Todos Santos went on to conduct a detailed hydrographic study of the streambed and lagoon at their doorstep and saw it through to publication (Eckman 2018).

VIEW FROM THE PACIFIC RIM

My former student is a wizard at mining the riches of the internet to link his classroom with the wider world. It is an antidote to the inherent isolation of a classroom in a small town at the end of a long peninsula. Knowing that the students sit within a stone's throw from the invisible line at the Tropic of Cancer (map 10), I asked if anyone could tell me what the first landfall might be if that line were followed west across the Pacific Ocean. No one was willing to hazard a guess. On reflection that's not surprising, because the Pacific Ocean is the planet's largest, and many world maps put the smaller Atlantic Ocean dead center with Europe and Africa on one side and the Americas on the other. In such a presentation, the Pacific Ocean is arbitrarily split down the center, often with New Zealand and the tail of Alaska's Aleutian Islands replicated at the margins like hypothetical bookends. With regard to terra firma, there exists much empty space on a map centered on this greatest of oceans. For those spending any time in Baja California, a map showing the full Pacific basin is required to grasp the interrelationships of physical geography, marine geology, and climate on a larger scale relevant to the peninsula (figure 10.3).

Lands that encircle the Pacific Ocean often are referred to as forming a ring of fire. Indeed, all of the Aleutian Islands, Japan, New Guinea, and most of New Zealand, as well as the totality of western South and Central America, are studded with a line of active volcanos. These are plumbed to magma chambers that align with the convergent boundaries between tectonic plates, of which the Pacific Plate is the world's largest. Convergent boundaries are places where ocean crust is bent and subducted beneath continental crust and where at depth the crust is melted and recycled. Based on the occurrence of earthquakes associated with plate convergence, it is even possible to configure the three-dimensional shape of crustal descent, typically at 45° bent beneath the leading edge of an opposing plate. Ongoing debate among geophysicists questions whether or not the immense weight of basaltic slabs contributes to their sinking at oceanic trenches. But there is no doubt that oceanic crust is pushed laterally to the sides of divergent plate boundaries, where magma rises to form new ocean crust in submarine rift valleys. The East Pacific Rise is the most commanding seafloor feature in the Pacific Ocean (figure 10.3), where rates of crustal spreading are among the highest on the planet. Fresh magma entering rifts from below adds as much 12 in (30 cm) of new crust per year, half of which is wedged aside in one direction and half in the opposite direction by yet more magma.

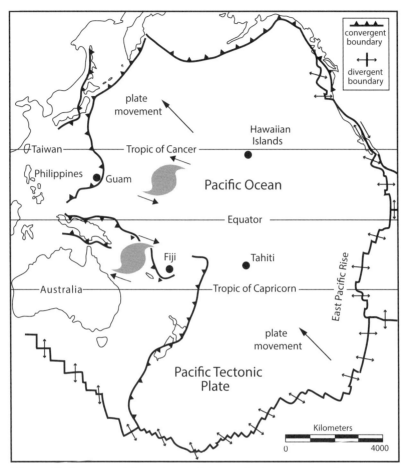

Figure 10.3. Map with boundaries of the Pacific Tectonic Plate shown against the greater Pacific Ocean basin and schematics for typhoon storms in a counterclockwise rotation in the northern hemisphere and clockwise rotation in the southern hemisphere.

For a student growing up in Todos Santos, the astonishing fact about the East Pacific Rise is its meeting with the Gulf of California. Essentially, an actively divergent boundary has been caught beneath the leading edge of continental Mexico, and resulted in the detachment of the Baja California peninsula as a geographic sliver pulled from the rest of the mainland. Prior to about 12 million years ago, a normal convergent boundary extended between present-day Cabo San Lucas in Baja California and San Francisco in Alta California through a line of volcanos connected to deep chambers where magma associated with granite accumulated and cooled over time (Lyle and Ness 1991). The granite mountains of the Sierra de la Laguna at the tip of the Baja California peninsula, as well as the Laguna Mountains near San Diego in Alta California, are relics from that time, now unroofed by erosion. With detachment, the volcanic systems became unplugged. Due to the interplay of the East Pacific Rise with the rest of North America, the Californias no longer belong to the North American Tectonic Plate, but now ride on the Pacific Plate. It is noteworthy that many other volcanos in the middle of the Pacific Plate, such as those in the Hawaiian Islands and Tahiti (figure 10.3) are unrelated to the dynamics of convergent boundaries. They are, instead, correlated by geophysicists with random plumes of magma that rise through oceanic crust from deep within the earth's mantle like a welder's torch.

Returning to the Tropic of Cancer, the task remains to identify a sister town to Todos Santos on the other side of the Pacific Ocean. Technically, the first landfall due west from Todos Santos is tiny Necker Island (also known as Mokumanamana) located a little more than 300 miles (~500 km) northwest of Kauai in the Hawaiian island chain. It is an uninhabited island, only 45 acres (18 ha) in size, but of significance to native Hawaiian peoples who visited there based on archaeologic evidence.

Leaving the Pacific Plate, but continuing across the greater Pacific Ocean, the first major landfall on the Tropic of Cancer is the island of Taiwan, off the coast of mainland China. It is a large island, encompassing 13,974 square miles (36,192 km²) and home to nearly 24 million residents. Overall, Taiwan is a third the size of the Baja California peninsula, but has a resident population nearly six times that of Baja California. The town of Ch'engkung sits just below the Tropic of Cancer on the east coast of Taiwan and makes a suitable candidate for a sister town to Todos Santos. Since 2010, more than 20 typhoons have made landfall on the outer shores of Taiwan. Super Typhoon Lekima struck in August 2019, considered the most intense since 2014. The schoolchildren of Ch'engkunk surely know about storm damage associated with Pacific tropical disturbances.

Taiwan is separated from mainland China by a strait only 81 miles (130 km) wide at its narrowest. Mainland China's coastline stretches

8,700 miles (14,000 km) in length, virtually all of which is vulnerable to storm damage. The country has in place an extensive system of coastal barriers along 85 percent of its shore, designed to protect against storm intensity ranging between 50- and 100-year events.[5] Heavy rainfall and landslides affected 5 million people in Zhejiang Province, when the same Super Typhoon Lekima departed Taiwan and struck the mainland. Meteorologists reckon it was among the three strongest typhoons ever to have impacted the province. Farther north (figure 10.3), Typhoon Hagibis hit Japan in October 2019 as one of the strongest in recent years, but was downgraded to a Category 2 event by the time it reached Tokyo, with a maximum wind speed of 69 mph (111 km/hr). Up to 3 ft (90 cm) of rain fell in 24 hours, a record for the area that caused widespread flooding.

Midway between the Tropic of Cancer and the equator, the Philippine Islands (figure 10.3) are often impacted by major storms, none more violent than Super Typhoon Yolanda (also identified as Typhoon Haiyan in neighboring east Asia). Yolanda made landfall on Leyte Island, southeast of the capital Manila, in November 2013, packing peak wind speeds of 195 mph (315 km/hr). Coastal surge from wind-driven waves having a maximum height of 60 ft (~18 m) did most of the damage to infrastructure and caused more than 6,000 fatalities. Based on wind speed alone, meteorologists consider Yolanda to have been one of the strongest tropical disturbances ever to make landfall. Two years later, a team of geomorphologists visited Calicoan Island close by the storm track to measure the enormity of limestone blocks peeled from coastal cliffs up to 40 ft (~12 m) in height and subsequently transported as much as 920 ft (280 m) inland. Blocks weighing between 50 and 180 metric tons were found to be flipped over by storm surge (Kennedy et al. 2017). Aspects of this survey set the standard for the kind of studies done at Ensenada Almeja (see chapter 4), Arroyo Blanco on Isla del Carmen (see chapter 5), and Puerto Escondido (see chapter 6) in Baja California Sur. The loss of life suffered in the Philippines was equivalent to wiping out the entire town of Todos Santos by a comparable hurricane.

No overview of the Pacific basin is complete without consideration of Australia, which is bisected by the Tropic of Capricorn in the Southern Hemisphere (figure 10.3). Cyclones are a common phenomenon in the Coral Sea east of Queensland, but they rotate clockwise, which is the opposite of typhoons or hurricanes in the northern hemisphere. The 2018/19 storm season for Australia was busy with the arrival of 11 events, 6 of which intensified as severe tropical disturbances. Cyclone Debbie in March 2017 was among the strongest on record to hit Queensland, making landfall as a Category 4 event near Townsville with maximum gusts of 155 mph (250 km/hr). Coastal damage was due mostly to storm surge, although torrential rains that exceeded 29 in (1 m) in 48 hours affected larger parts of Queensland. The first Category 5 storm of the 2020 season left the Solomon Islands in the Coral Sea northeast of Australia and followed a course making a direct hit on Espiritu Santo in the Vanuatu Islands on April 5 with winds packing 170 mph (275 km/hr). The island suffered extensive defoliation of trees from the winds and massive flooding from 18 in (45 cm) of rain that left thousands of inhabitants homeless. The cyclone lost strength before skirting the Fiji Islands farther to the southeast (figure 10.3).

Whether called hurricanes, typhoons, or cyclones, or whether they rotate counter-clockwise or clockwise, all such storms are spawned by excessively warm surface ocean water. Global warming between the Tropics of Cancer and Capricorn continues to fuel the recurrence and duration of elevated sea-surface temperatures within those latitudes. As cited in chapter 1, evidence from global data suggests that pools of warm surface water are forming more frequently and lasting longer during the passage of years since 1925 (Oliver et al. 2018). Based on global data collected since 1955, it is now calculated that the five warmest years in the context of warmth retained by ocean water span from 2015 to 2019, with each succeeding year setting a new record (Cheng et al. 2020).

An Epilogue

Global warming is no hoax. Based on detailed climate records, the intensity of tropical storms fed by warming ocean-surface waters is shown to have increased by 8 percent per decade over the last 40 years (Kossin et al. 2020). It behooves us as a species uniquely capable of listening to our planet and learning from its story to come to terms with our role as an agent of major change. True enough, Planet Earth endured fluctuations in sea level and climate change far more extreme in the past than anything now in front of us. Our species emerged no more than 300,000 years ago and cannot be blamed for any of the climate-related mass extinctions that took place during the last 500 million years. What is different today is the extraordinarily rapid rise in atmospheric carbon dioxide as the chief incubator of rising global temperature both on land and at sea. Burning of fossil fuels, mainly coal at the beginning of the Industrial Revolution in the late eighteenth century, started the rise in carbon dioxide when the world population was barely 700 million. Today, we are 7.85 billion people, and the amount of carbon dioxide in the atmosphere has spiked with the explosion of our numbers, alarmingly so on a decadal basis. Climate cycles that predate us humans modulated much more slowly, with the latest glacial and interglacial cycles separated by 100,000 years on average.

Reading the rocks, including those forming before our eyes like the boulder barriers at Puerto Escondido, is a skill available to anyone who takes the time to go out in nature. It is a joyful thing to introduce that kind of literacy to young and old alike in such a magnificent setting as the Baja California peninsula. It is something to revel in, even more so during a time of extraordinary challenge. The students of Todos Santos and Taiwan's Ch'engkunk are adaptable and eager to learn. We grown-ups cannot afford their sacrifice to the perpetuation of old economies. How can we all come to terms with a future that portends such stark changes in climate? We have no choice but to join hands across the oceans that separate us, to practice new habits in the elimination of waste in food and material goods, and to maximize the use of renewable sources of energy to the fullest extent possible. Some would say that a new species of altered humans is on the horizon by the century's end. Let it not be in a dystopian world, but one in which the challenge of a sustainable economy is the common goal.

Glossary of Geologic and Ecologic Terms

ANDESITE. Igneous rocks from surface flows (volcanic lavas and breccia) rich in the minerals plagioclase and lesser amounts of pyroxene, hornblende, and/or biotite. Although a freshly broken surface is generally dark gray in color, weathered surfaces often take on a reddish cast due to oxidation of iron content. The name is derived from the Andes, which is representative of the mountain belt where these rocks occur due to ocean-plate subduction against a continental margin.

ANTICLINE. A structural deformation of the earth's crust in which rock layers are compressed to make a positive fold with sides that slope downward away from a central fold axis.

ARCHAEA. Prokaryotes (single-cell organisms) that reproduce asexually and live under anaerobic conditions, unlike bacteria that thrive under aerobic conditions.

BARNACLE. Marine invertebrate, classified within the phylum Arthropoda, solitary in habit, attached to rocks via a cemented basal plate surrounded by flexible plates configured in a dome shape. Feeding appendages (cirri) are extruded into the water column to sweep up food particles in suspension.

BASALT. Igneous rocks of an extrusive origin (flood lavas or submarine pillow lavas) rich in the minerals plagioclase, pyroxene, and often olivine. These rocks typically form on the ocean floor or in continental rifts. They are dark and generally weather dark.

BASEMENT ROCKS. The intrusive igneous and/or metamorphic rocks overlain at depth by sedimentary rocks.

BATHYMETRY. Measurement of depth of water in oceans, seas, or lakes that results in a bathymetric map using lines the same way as in a topographic map.

BEDROCK. Solid rock, stratified or unstratified, that occurs under soil, gravel, or any other unconsolidated surface materials.

BERM. A well-demarcated deposit of sand or gravel that is level in extent and marks the maximum landward extent of reworking by waves. A beach may have more than one berm, depending on the effect of tidal cycles and storms.

BRAIDED STREAM. A river with a streambed divided into multiple dividing and reuniting channels, looking superficially like the woven stands of a braid. This morphology is most often the result of more water being introduced to the stream than can be carried downstream.

BRYOZOAN. Invertebrate, classified within the phylum Bryozoa, colonial in habit, mostly marine but some dwell in freshwater. A feeding appendage (lophophore) sits within a calcified cup and is extruded into the water column to collect food particles in suspension.

CARBONATE RAMP. A gently sloping surface (generally 5° to 10°) that forms as a continuum from shallow to deeper water and consists of carbonate sediments when under active construction, or limestone when cemented in place. The ramp may sit on an unconformity surface eroded from preexisting rocks (sedimentary, igneous, or metamorphic) by coastal and nearshore processes.

CATASTROPHISM. The "new" concept of catastrophism (in contrast to the late eighteenth-century concept of events outside the normal laws of nature) invokes violent hurricane and tsunami events that do not occur on a daily basis.

CHUBASCO. Spanish term adapted by Mexicans for a major rainstorm, but in Spain the term refers to an ordinary shower.

CLAST. An individual fragment of rock (varying sizes), eroded by the action of wind, waves, or running water from a parent source.

CLASTIC ROCKS. Layered sedimentary rocks that include conglomerate, sandstone, shale, and siltstone.

COASTAL BOULDER DEPOSIT. Boulders are defined by convention as individual clasts equal to or larger than 10.1 in (25.6 cm) in diameter. There is no defined upper limit to the size of a boulder. A deposit of boulders that sits on or adjacent to a rocky shoreline may owe its origin to a major storm (or succession of storms) that loosened the rocks from their point of origin in a cliff line or from a tsunami event that accomplished the same result.

COASTAL OUTWASH DEPOSIT. A combination of boulders and smaller clasts, including cobbles and pebbles, may form a coastal deposit as a result of heavy rainfall due to a major storm (or succession of storms) that loosens and washes the clasts to the shore from an inland source.

CONGLOMERATE. Sedimentary rocks consisting of cemented clasts (pebbles, cobbles, boulders) with little intervening matrix.

COQUINA. Bedded accumulations of cemented shells that typically exclude any intermittent sediment.

CORAL. Marine invertebrates that are solitary or colonial and belong to the phylum Coelenterata recognized by paleontologists (phylum Cnidaria to biologists). Many species secrete a solid skeleton ($CaCO_3$) and are the principal contributors to the construction of reefs.

CORALLINE RED ALGAE. Marine plants that belong to the division Rhodophyta and have the ability to secrete skeletons of calcium carbonate. The adjective "coralline" is applied to indicate that the algae mimic corals in appearance.

DEBRIS FLOW. A kind of landslide consisting of rocks, soil, and mud in which more than half of the clasts are larger than sand-size particles.

DIKE. A tabular body of igneous rock injected as a magma through cracks and fissures in a preexisting body of rock. When softer rocks are eroded on opposite sides of the dike, it takes on the appearance of a freestanding wall.

ECHINODERM. Marine invertebrates that are solitary in plan, exhibit five-fold symmetry, and belong to the phylum Echinodermata. Many species secrete a shell (test) formed by $CaCO_3$. Common examples include sea urchins, sand dollars, starfish, and sea cucumbers.

EL NIÑO EVENT. Regional climate change localized between the Tropic of Cancer and Capricorn that results in major oceanic storms is stimulated by the buildup of unusually warm sea-surface temperatures over the equator, especially the enormous stretch of the equatorial Pacific Ocean. An event that normally lasts for several months to a year or more also involves a slowing in the equatorial currents that move surface water from east to west. Cold-water currents like the Humboldt Current off the Pacific coast of South America also are disrupted by this slowdown. The original phrase in Spanish, *El Niño de Navidad,* was coined by Peruvian fishermen who related episodic warming events to the newborn Christ at Christmas time.

ESCARPMENT. A steep rock face showing the termination of stratified rocks that dip somewhat into the outcrop.

EXTENSIONAL STRESS. Force applied to the earth's crust that results in stretching or pulling in opposite directions.

FACIES. Sedimentary rocks and fossils that represent contemporaneous variations in a lateral continuum. A common representation relates to facies changes in an onshore-offshore pattern.

FAUNA. The animals that live in a given area or environment. A faunal list gives the names of those animals (or fossils) found in a given habitat.

FORAMINIFERA. Marine invertebrates that consist of a single cell but characterized by an exoskeleton called a test that may precipitated by the animal ($CaCO_3$) or constructed by the animal from an inorganic source (SiO_2). Many are planktonic, but others live in a neutrally buoyant state at different water depths. Others dwell on the seafloor.

GEOMORPHOLOGY. The study of physical landforms and the natural processes that lead to their development at the surface of the earth.

GRABEN. An elongated fault block that is downthrown with respect to adjacent blocks.

GRANITE. Igneous rock that cooled far underground with large mineral crystals that typically include feldspar, plagioclase, quartz, and biotite.

HOLOCENE. Most recent geologic time, measured by convention as the last 10,000 years.

HORST. An elongated fault block that is upthrown with respect to adjacent blocks on either side. The effect is to create a crested topography.

HURRICANE. Name applied to tropical storms with heavy rainfall in the North Atlantic Ocean and eastern North Pacific Ocean accompanied by strong winds that circulate in a rapidly rotating system characterized by low atmospheric pressure at the center.

HYALOCLASTITE. A volcanoclastic accumulation or kind of breccia consisting of clasts with high silica content (SiO_2) formed during the sudden chilling of a submarine lava flow on entry into seawater. The term is derived from the Greek word *hyalus*, for glass.

IGNEOUS ROCKS. Rocks cooled from molten material either deep within the earth's crust (intrusive) or at the earth's surface (extrusive) as a result of volcanic activity.

IGNIMBRITE. Igneous rock deposit formed by volcanic action in which ash and other material is conveyed downslope at great speed cushioned from below by trapped gas.

JOINT. A vertical fracture in rocks.

KARST. A range of landforms that develop both on and within terrain dominated by limestone cover due to the dissolution of $CaCO_3$ under a humid climate. The name comes from the Karst district on the coast of the Adriatic Sea.

KRUMMHOLZ. Term derived from German (twisted wood) that applies to trees or shrubs contorted from normal upright growth to a bent or even ground-hugging posture due to the deleterious effects of salt, sand, or ice crystals carried by the wind.

LIMESTONE. Sedimentary rock consisting of calcium carbonate ($CaCO_3$) derived primarily from organic remains of marine invertebrates such as corals, mollusks, echinoderms, and coralline algae.

MARINE TERRACE. A narrow coastal rim that usually slopes gently seaward and is veneered by a marine deposit. Formation of the terrace is caused by intertidal erosion, and the position of the terrace depends on changes in global sea level with respect to changes in the local or regional elevation of the coastline.

MICROBIAL COMMUNITY. A naturally occurring assemblage of bacterial and/or archaeal species that often thrive under extreme conditions such as very high salinity or very high temperatures to the exclusion of other species of plants and animals.

MIOCENE. A geologic epoch, roughly 18 million years in duration, that began about 23 million years ago and terminated a little more than 5 million years ago. All those sedimentary and igneous rocks that originated during that interval are said to belong to the Miocene Series.

MOLLUSK. Marine invertebrates that are solitary in plan and belong to the phylum Mollusca. The phylum includes land and sea snails (class Gastropoda), clams (class Bivalvia), as well as squids and the octopus (class Cephalopoda). In particular, the shelled gastropods and bivalves (such as pectens) lend themselves to fossilization.

MUDBALL. A spherical congealing of sticky mud that includes shells and/or other rock debris picked up during a submarine gravity slide, which can be triggered by an earthquake.

MUDFLOW. A kind of landslide characterized by a water-saturated mass dominated by small particles, such as silt and clay. The flow is capable of moving large objects such as boulders (or even houses) that tend to float within the mass.

OBSIDIAN. Solid volcanic glass (SiO_2), often black. Typically forms a conchoidal fracture when broken.

PLEISTOCENE. A short geologic epoch, dating from about 2,588,000 years ago and ending about 10,000 years ago, that bridges the prior Pliocene Epoch and the Holocene (Recent). All sedimentary and igneous rocks that originated during that time interval are said to belong to the Pleistocene Series.

PLIOCENE. A geologic epoch, roughly 2.8 million years in duration, that began more than 5 million years ago and terminated about 2,588,000 years ago. All those sedimentary and igneous rocks that originated during that interval are said to belong to the Pliocene Series.

PLUTON. An intrusive body of igneous rock formed beneath the surface from slow-cooling magma. Granite, diorite, and tonalite are examples.

PLUVIAL LAKE. A lake that formed during a geologic interval when rainfall was locally more abundant than today. Present desert regions in the northern hemisphere typically were subjected to higher rates of rainfall during the various Pleistocene glaciations. When desert conditions returned during the interglacials, the old lakebeds and lake terraces were exposed.

PUMICE. Volcanic glass with a froth-like texture formed by open cells that trap air. The material is extremely light and individual pieces will float in water.

RANGE ZONE. The first and last occurrences of a given species of fossil traced through a continuous succession of sedimentary strata delimits that species' documented range, which through the correlation of index fossils also relates to the passage of geologic time.

RHODOLITH. A particular kind of coralline red algae that grows unattached on the seafloor. The rhodolith assumes a spherical shape due to frequent movement with wave and current activity during the lifetime of the alga. The alga may colonize a tiny piece of shell or a rock fragment as large as a pebble, thereafter growing outward in a radial pattern.

RHYOLITE. A volcanic rock formed as a surface flow that is chemically the fine-grained equivalent of granite.

RIFT ZONE. A region, typically linear in demarcation, where a continent has begun to break apart or where ocean crust continues to spread apart in opposite directions.

ROCK PEDIMENT. A sizable knob of rock that projects above a connecting pillar of rock forms due to differential erosion. Also called a mushroom rock on account of the shape.

SAFFIR-SIMPSON HURRICANE WIND SCALE. The categorization of hurricanes (also typhoons) based on a rating from 1 to 5, where the base line for a Category 1 storm is achieved with a wind speed of 74 mph (119 km/hr) and a Category 5 storm is achieved with a wind speed of 157 mph (252 km/hr).

SEA ARCH. Coastal rock formation that results from wave erosion that tunnels into sea cliffs from opposite sides of a protruding headland. A pair of sea caves is the precursor to a sea arch, and once the arch collapses from ongoing erosion, the result is a sea stack.

SEDIMENTARY ROCKS. Rocks formed by the burial and cementation of inorganic sediments like pebbles, sand, silt, and clay, or the fragments of broken corals and shells that form limestone.

SPECIFIC GRAVITY. The ratio of the weight of a given mineral to the weight of the same volume of water. Galena (lead) has a specific gravity of about 7.5, while quartz (SiO_2) has a specific gravity of only 2.6.

STRATA. Layers of sedimentary rocks.

STRATOVOLCANO. Steep-sided volcanos often high in elevation that add layers of ash and other ejecta to the sides in a cone-shaped profile. The Italian island of Stromboli is the ideal example for this class of volcanos.

STRATUM. A single layer (or bed) in a sequence of layered sedimentary rocks.

STRIKE-SLIP FAULT. A fault in which the net slippage is confined to the direction of the fault strike. That is, movement on opposite sides of the fault trace is seen mainly as lateral, as opposed to vertical. Common synonyms are wrench fault and transcurrent fault.

STROMATOLITE. A simple microbial deposit made by cyanobacteria and other microbes, typically laminar in organization, originating far back in Precambrian time. The name derives from the Greek *stromat*, meaning to spread out; Latin *stroma*, for bed; and "lithos" for stone.

SUPERPOSITION. In conformity to the concept of *original horizontality*, the interpretation that the bottom layer in a sequence of stratified rocks is the oldest bed and the top layer is the youngest bed.

THERMOCLINE. Changes in oceanic water temperature that rapidly decline from the surface with depth until reaching a level after which the change is much more gradual.

TRANSFORM FAULT. A major fracture that runs perpendicular to an ocean ridge and along which strike-slip movement occurs.

TROPICAL CYCLONE. General term that applies to hurricanes and typhoons that swirl in rotation.

TUFF. A rock formed from volcanic ash and small fragments (usually less than 4 mm, or 1/8 in, diameter) of volcanic rock blasted by an eruption.

TUFFACEOUS. Adjective for sediments more than 50 percent composed of tuff.

TYPHOON. Term applied to tropical storms with heavy rainfall in the western North Pacific Ocean accompanied by strong winds that circulate in a rapidly rotating system characterized by low atmospheric pressure at the center. Essentially the same as a hurricane.

UNCONFORMITY. A surface of erosion that separates two bodies of rock and represents an interval of time during which deposition ceased, some material was removed, and then deposition resumed again. An *angular unconformity* involves two sets of stratified rocks on opposite sides of the unconformity surface, but other types of unconformities may involve a juncture between sedimentary rocks and igneous or metamorphic rocks.

UNIFORMITARIANISM. The basic concept that the same physical processes that shaped the earth throughout geologic time in the past are the same processes we may observe in action today.

WAVE-CUT NOTCH. Horizontal cleft eroded in a rocky shoreline by persistent wave action generally limited to the tidal zone.

Notes

Chapter 1

1. The Baja California peninsula hosts a rich biota of cacti and other unique plants characteristic of an arid setting receiving a mean average rainfall of only 6 in. (15.3 cm). The flora is described in the third edition of the *Baja California Plant Field Guide*, by Jon P. Redman and Norman C. Roberts (San Diego: San Diego Natural History Museum and Sunbelt Publications, 2012). The term *chubasco* is introduced therein (p. 6) in regard to intense rainfall related to major storms between droughts lasting as long as six years.

2. A review of the Pleistocene coral reefs in Baja California Sur and its associated islands in the Gulf of California is provided by Johnson, M. E., R. A. López-Pérz, C. R. Ranson, and J. Ledesma-Vázquez, "Late Pleistocene Coral Reef Development on Isla Coronados, Gulf of California," *Ciencias Marinas* 33 (2007): 105–120.

3. The rhodolith bed in the channel between Isla Coronados and the peninsular mainland at El Bajo was documented for the first time in a contribution by M. S. Foster, R. Riosmenta-Rodrigues, D. L. Steller, and W. J. Woelkerling, "Living Rhodolith Beds in the Gulf of California and Their Implications for Paleoenvironmental Interpretation," in *Pliocene Carbonates and Related Facies Flanking the Gulf of California, Baja California, Mexico*, ed. M. E. Johnson and J. Ledesma-Vázquez, 127–39, Geological Society of America Special Paper 318 (Boulder, Colo.: n.p.), 2007.

4. The story of the Isla Cerralvo expeditions is told in chapter 9 of "Riding Out Ancient Storms on Isla Cerralvo," in M. E. Johnson, *Off-Trail Adventures in Baja California* (Tucson: University of Arizona Press, 2014).

5. Online list of Baja California Peninsula hurricanes by Wikipedia: https://en.wikipedia.org/wiki/Wik/List_of_Baja_California_Peninsula_hurricanes.

6. Online access to the NASA mission for Global Precipitation Measurement (GPM): https://www.nasa.gov/mission_pages/GPM/main/index.html.

7. See coverage by D. H. Backus and M. E. Johnson in "Sand Dunes on Peninsular and Island Shores in the Gulf of California," in *Atlas of Coastal Ecosystems in the Western Gulf of California*, ed. M. E. Johnson and J. Ledesma-Vázquez, 117–33 (Tucson: University of Arizona Press, 2009).

Chapter 2

1.   The authoritative geological map for the northern state of Baja California remains the three sheets published in 1971 compiled by R. Gordon Gastil, Richard P. Phillips, and Edwin C. Allison, published by the Geological Society of America.

2.   An article in *National Geographic* from May 1984 (vol. 165, no. 5, 556–613) under the title "The Dead Do Tell Tales at Vesuvius" by Rick Gore relates the most evocative description of a town's almost instantaneous death struggle against an ignimbrite that buried Herculaneum near Pompeii. The piece portrays in striking images the scene where some 150 human skeletons were excavated, clustered together at the town's shore, where hot volcanic ash and gasses caused death by asphyxiation.

3.   The Pedra de Lume salt works inside the caldera of an extinct volcano on Sal Island off the coast of West Africa and is described in the guidebook *Cape Verde Islands*, by A. Irwin and C. Wilson (Chalfont St. Peter, UK: Bradt Travel Guides, 2009), 113. The volcano is two-thirds of a mile (1 km) from the ocean, and seawater enters the caldera by natural seepage through the porous volcanic rocks.

4.   The discovery story regarding the living stromatolites on Isla Angel de la Guarda is told in chapter 2 of M. E. Johnson, *Off-Trail Adventures in Baja California* (Tucson: University of Arizona Press, 2014).

5.   Refer to the Annual Mean Precipitation chart for the Baja California peninsula (p. 5) in the third edition of the *Baja California Plant Field Guide*, by Jon P. Redman and Norman C. Roberts (San Diego: San Diego Natural History Museum and Sunbelt Publications, 2012).

6.   The shell characteristics of *Chione californiensis* and its preferred habitat are described by Richard Brusca in his volume on the *Common Intertidal Invertebrates of the Gulf of California*, 2nd ed. (Tucson: University of Arizona Press, 1980), 146–49.

7.   Fossil shells including those of *Chione californiensis* are described in the author's hikes around Punta Chivato in *Discovering the Geology of Baja California* (2002) and Isla Coronados in *Off-Trail Adventures in Baja California* (2014) published by the University of Arizona Press.

8.   Third edition of the *Baja California Plant Field Guide* (2012), 8.

Chapter 3

1.   Posted on May 28, 2019, at *DieselNet News* (https://dieselnet.com/news/2019/05co2.php): "Atmospheric carbon dioxide levels hit another record high—CO_2 concentrations reached 415 ppm this month for the first time in human history according to data by the US National Oceanic and Atmospheric

Administration's (NOAA) Mauna Loa Observatory in Hawaii; the last time CO_2 concentrations were at 415 ppm was likely close to 3 million years ago."

2. The fossil species of sand dollar, *Dendraster granti*, is described in the treatise by J. Wyatt Durham under the title "Megascopic Paleontology and Marine Stratigraphy," published in the *Geological Society of America Memoir 42* issued in 1950 as the "1940 E.W. Scripps Cruise to the Gulf of California." Therein (pp. 41–42), the extinct species is assigned to a middle Pliocene range.

3. The satellite image dating from 2003 acquired from the NASA LAND-SAT Program appears as figure 3.7 on p. 37 in the *Atlas of Coastal Ecosystems in the Western Gulf of California*, edited by M. E. Johnson and J. Ledesma-Vázquez (Tucson: University of Arizona Press, 2009).

4. The chapter on "San Francisquito's Ancient Bay" (pp. 65–90) describes an ambitious day-long hike around the Pliocene Ensenada Blanco in M. E. Johnson, *Off-Trail Adventures in Baja California* (Tucson: University of Arizona Press, 2014).

Chapter 4

1. The extraordinary coastal trek around the Baja California peninsula by Graham Mackintosh is described in his 1988 book *Into a Desert Place* (London: Urwin Hayman). Another impressive feat of physical endurance was the voyage by kayak taken by Jonathan Waterman and his partner as told in his 1995 book *Kayaking the Vermillion Sea* (New York: Simon & Schuster). Both authors must have passed by Ensenada San Basilio, but make no mention of the encounter.

2. A technical report of the El Mangle area including its spectacular paleo-shoreline is covered in the paper authored by M. E. Johnson, D. H. Backus, and J. Ledesma-Vázquez under the title "Offset of Pliocene Ramp Facies at El Mangle by El Coloradito Fault, Baja California Sur: Implications for Transtensional Tectonics," published in 2003 as a contribution to *Geological Society of America Special Paper 374*, pp. 407–420. The popular account of this project is given in Chapter 6: Intersection of Fractures at El Mangle in *Off-Trail Adventures in Baja California* (Tucson: University of Arizona Press, 2014).

3. Some of the background regarding the contessa's house at San Basilio is outlined in a piece written by Paula Brook titled "Writing San Basilio," published as part of a group project initiated by the Loreto Writers Workshop in their privately published 2016 book *Reflections by the Sea* (ISBN 978-1-36-499242-2).

4. The vast landscape of the Cerro Mercenarios volcanic complex that impinges on San Basilio Bay appears as a massive volcano with a distinctive pattern of radial stream erosion in the aerial photo (plate 10) from *Off-Trail Adventures in Baja California* (2014).

5.   The phi scale for sedimentary grain sizes attached to figure 4.4 extends in this specific case from values pegged at 6 for medium silt (0.0156 mm in diameter) to -6 for cobbles (64 mm and larger in diameter).

6.   Mud balls formed during submarine slides that incorporate fossil pecten shells are described and illustrated in figure 9.7 on p. 115 from chapter 9 on "Growth of Pliocene-Pleistocene Clam Banks (Mollusca, Bivalvia) and Related Tectonic Constraints" in the *Atlas of Coastal Ecosystems in the Western Gulf of California*, ed. M. E. Johnson and J. Ledesma-Vázquez (Tucson: University of Arizona Press, 2009).

7.   Video shot by Erik Stevens on September 15, 2014, from a front door at the Spanish contessa's house may be viewed online in connection with the published article by M.E. Johnson, Guardado-France, E. M. Johnson, and Ledesma-Vázquez (2019) at the following address: http://www.mdpi.com/2077-1312/7/6/193/s1.

8.   *Oasis de Piedra*, with photographs by Miguel Ángel de Cueva (San Diego: Planeta Publishers, 2008), includes an aerial image with the following caption (translated from Spanish, p. 37): "Volcanic ash, frozen in broken stone, yields to swirls of sand where solids and liquids are united in Puerto Almeja."

9.   The dynamics behind the shell beach at Ensenada el Muerto are described in chapter 5 of *Discovering the Geology of Baja California* (Tucson: University of Arizona Press, 2002). Likewise, the comparable shell beach on Isla Monserrat is mentioned in chapter 8 of *Off-Trail Adventures in Baja California* (Tucson: University of Arizona Press, 2014), both by M. E. Johnson.

10.   Lithified sand dunes from the last interglacial epoch approximately 125,000 years ago are described from various localities in chapter 11 on "Beach Deflation and Accrual of Pliocene-Pleistocene Coastal Dunes of the Gulf of California Region," in *Atlas of Coastal Ecosystems in the Western Gulf of California*, ed. M. E. Johnson and J. Ledesma-Vázquez (Tucson: University of Arizona Press, 2009).

11.   The geographic range and specific habitat of *Periglypta multicostata* is described by M. A. Keen in *Sea Shells of Tropical West America*, 2nd ed. (Stanford: Stanford University Press, 1971). See page 161.

Chapter 5

1.   The 278-year history of salt extraction on Isla del Carmen is related in chapter 5 (pp. 173–97) of the book by Ann and Don O'Neil, *Loreto, Baja California: First Mission and Capital of Spanish California* (Studio City, Calif.: Tio Press, 2001).

2.   The earliest geological map of Isla del Carmen was compiled by the geologist Charles A. Anderson, who visited in 1940 as a member of the E. W. Scripps Cruise to the Gulf of California organized from the Scripps Institute of

Oceanography. The official report of the expedition, which logged 4,600 nautical miles (8,500 km) of travel during the 65 days spent on exploration within the Gulf of California, includes only a brief mention of thick "volcanic gravels" that sit on older igneous rocks correctly classified as andesite belonging to the Miocene Comondú Group. Anderson's report is contained in part 1, Geology of Islands and Neighboring Land Areas, 1940 E.W. Scripps Cruise to the Gulf of California, published in 1950 as *Geological Society of America Memoir 43*.

    3.   A photo on page 91 from the chapter on limestone generated by rhodoliths captures the rose-tinted cliffs that faithfully reflect their origin from the eroded debris of coralline red algae. Chapter 7 by M. E. Johnson, Backus, and Riosmena-Rodríguez on "Contribution of Rhodoliths to the Generation of Pliocene-Pleistocene Limestone in the Gulf of California," in the *Atlas of Coastal Ecosystems in the Western Gulf of California*, ed. M. E. Johnson and J. Ledesma-Vázquez (Tucson: University of Arizona Press, 2009).

    4.   Taxonomy of living coralline red algae that secrete rhodoliths in the shallow waters off the south end of Isla del Carmen, as well as other neighboring areas, is treated in the article by R. Riosmena-Rodríguez, J. M. López-Calderón, E. Mariano-Meléndez, A. Sánchez-Rodríguez, and C. Fernández-Garcia, "Size and Distribution of Rhodolith Beds in the Loreto Marine Park: Their Role in Coastal Processes." *Journal of Coastal Research* 28 (2012): 255–60.

    5.   The habits and distribution of the extant "rock-eating" bivalve *Lithophaga aristata* is described on p. 136 in the *Common Intertidal Invertebrates of the Gulf of California*, 2nd ed. (Tucson: University of Arizona Press, 1980), by Richard C. Brusca.

    6.   At Punta Cacarizo close to the Punta Chivato promontory, salt precipitation occurs in small limestone cavities that may retain a thin film of seawater looking much like a stretched cellophane cover. An example is illustrated in figure 7 (page 45) in the guidebook *Discovering the Geology of Baja California* (Tucson: University of Arizona Press, 2002), by Markes E. Johnson.

    7.   The various tectonic stages through which the landscape of the Baja California peninsula has evolved, and continues to develop, are outlined in the introductory chapter to the *Atlas of Coastal Ecosystems in the Western Gulf of California* (Tucson: University of Arizona Press, 2009) by J. Ledesma-Vázquez, M. E. Johnson, O. Gonzale-Yajimovich, and E. Santamaría-del-Angel under the heading "Gulf of California Geography, Geological Origins, Oceanography, and Sedimentation Patterns."

    8.   Historical markers at the mission church of San Javier explain the connection with Francisco Javier (also written Xavier), the Basque native who in 1534 was among the first novitiates admitted to of the Society of Jesus. He is regarded as among the most prominent Roman Catholic missionaries, who under protection of the Portuguese King John II introduced Christianity to India, the Spice Islands of the Malay Archipelago, and Japan between 1542 and 1551. Two hundred and twenty-two years had passed after the death of the canonized San

Javier, when construction of the mission church in the high Sierra de la Giganta was undertaken. Additional markers provide details on the church's architecture and rich interior decoration.

9.   The geology of the Punta Perico conglomerate and faults associated with the east valley wall of the salt flats are described in the article by R. J. Dorsey, P. J. Umhoefer, J. C. Ingle, and L. Mayer from 2001 under the title "Late Miocene to Pliocene Stratigraphic Evolution of Northeast Carmen Island, Gulf of California: Implications for Oblique-Rifting Tectonics," *Sedimentary Geology* 144:97–123.

Chapter 6

1.   During of summer of 1923, Steinbeck registered for courses in general zoology as well as English composition offered at the Hopkins Marine Station affiliated with Stanford University, where he had completed his third year. The connection is described in *A Journey into Steinbeck's California*, by Susan Shillinglow, 2nd ed. (Berkeley, Calif.: A Roaring Forties Press Publication, 2006). See p. 104.

2.   The intertidal habitat and feeding preferences of the gastropods *Turbo fluctuosus* and *Murex elenensis* are described on pages 157 and 174 in the *Common Intertidal Invertebrates of the Gulf of California*, 2nd ed. (Tucson: University of Arizona Press, 1973), by Richard C. Brusca.

3.   The idea that sea level was about 20 ft (6 m) higher 125,000 years ago is premised on the elevation of abandoned reefs and former shorelines on islands thought to have remained tectonically stable over time. Two examples include Barbados and Bermuda. See M. L. Bender et al., "Uranium-Series Dating of the Pleistocene Reef Tracts of Barbados, West Indies," *Geological Society of America Bulletin* 90 (1979) 577–94; and A. C. Neumann and P. J. Hearty, "Rapid Sea-Level Changes at the Close of the Last Interglacial (Substage 5e) Recorded in Bahamian Island Geology," *Geology* 24 (1996): 775–78.

4.   An abundance of the data on marine terraces, including the distinctive 40 ft (12 m) terrace traceable around the Gulf of California, is available in the 1991 study by L. Orlieb on "Quaternary vertical movements along the coasts of Baja California and Sonora," in *The Gulf and Peninsular Province of the Californias*, ed. J. P. Dauphin and B. R. T. Simoneit, 447–80, American Association of Petroleum Geologists Memoir 47.

5.   In a chapter about Puerto Escondido from *Searching for Steinbeck's Sea of Cortez* (Seattle: Sasquatch Books, 2002), author Andromeda Romano-Lax comments erroneously that "most of this Range of the Giants is sedimentary rock, laid down when this whole region was covered by the sea" (p. 80). Indeed, the Salto Formation of the Comondú Group consists of sandstone exposed in Tabor Canyon, but those sands are largely eolian in origin from windblown

dunes. Note: Sierra de la Giganta translates as the Mountain of the Giant (singular, feminine).

Chapter 7

1.   Forty-three named islands are listed and ranked by size in the review article on the "Geology and Ages of the Islands," by Ana Luisa Carreño and Javier Helenes, in *A New Island Biogeography of the Sea of Cortés*, ed. T. J. Case, M. L. Cody, and E. Ezcurra, 114–40 (Oxford: Oxford University Press, 2002).

2.   The Taconic Range was first given its mantle of geological significance by one of my illustrious predecessors at Williams College, Professor Ebenezer Emmons, who first published on the Taconic System in his 1844 book: *The Taconic System Based on Observations in New York, Massachusetts, Main, Vermont and Rhode Island*, (Albany, N.Y.: Carroll & Cook).

3.   A dense concentration of wind-pruned trees (krummholz) at Punta Chivato is discussed and illustrated in *Discovering the Geology of Baja California*, by M. E. Johnson (Tucson: University of Arizona Press, 2002). Throughout the Gulf of California, such trees typically bend to the south in response to the seasonal winter winds from the north.

4.   A topographic map for Isla Monserrat showing a combination of both north to south and northwest to southeast trending faults is available for comparison in chapter 8 (Song of the Amazon on Isla Monserrat) in *Off-Trail Adventures in Baja California* (Tucson: University of Arizona Press, 2016), by M. E. Johnson.

Chapter 8

1.   The landscape around Cataviña and its surreal atmosphere is described in the opening chapter of *Off-Trail Adventures in Baja California*, by M. E. Johnson (Tucson: University of Arizona Press, 2016.)

2.   Among the forty-three named islands listed in the review article on the "Geology and Ages of the Islands" by Ana Luisa Carreño and Javier Helenes, in *A New Island Biogeography of the Sea of Cortés* (2002), 14–40, seven are attributed to granite.

3.   M. L. Bender et al., "Uranium-Series Dating of the Pleistocene Reef Tracts of Barbados, West Indies," *Geological Society of America Bulletin* 90 (1979) 577–94; and A. C. Neumann and P. J. Hearty, "Rapid Sea-Level Changes at the Close of the Last Interglacial (Substage 5e) Recorded in Bahamian Island Geology," *Geology* 24 (1996): 775–78.

4.   L. Orlieb "Quaternary vertical movements along the coasts of Baja California and Sonora," in *The Gulf and Peninsular Province of the Californias*

(1991), ed. J. P. Dauphin and B. R. T. Simoneit, 447–80, American Association of Petroleum Geologists Memoir 47.

5. See the article by A. G. Coats et al., "Closure of the Isthmus of Panama: The Near-Shore Marine Record of Costa Rica and Western Panama," *Geological Society of America Bulletin* 104 (1992): 814–28.

6. See the chapter by Ted J. Case on "Reptile Ecology" in *A New Island Biogeography of the Sea of Cortés* (2002), 221–70, including an extensive, island-by-island master list (appendix 8.3, pp. 586–591).

7. See scientific correspondence from E. J. Censky, K. Hodge, and J. Dudley under the title "Over-Water Dispersal of Lizards Due to Hurricanes," *Nature* 395 (1998): 556.

8. The rhombic solution of Shepard (1950) is illustrated in his original publication, but also is applied as an explanation of local tectonics at Punta Chivato by M. E. Johnson in chapter 7 on Islas Santa Inés (pp. 149–75) in *Discovering the Geology of Baja California* (Tucson: University of Arizona Press, 2002).

Chapter 9

1. Availability of snorkeling equipment would not have been useful to Ed Ricketts, who never learned to swim.

2. In Steinbeck's annotated account of the 1940 excursion aboard the *Western Flyer*, he details the experiment with the coral fronds at Cabo Pulmo (1951, 78), and states that the "thousands" of hidden inhabitants were dominated by small marine worms and tiny crabs.

3. Refer to notes on habitat from the *Common Intertidal Invertebrates of the Gulf of California*, 2nd ed., by Richard C. Brusca, on p. 164 for *Strombus galeatus* and p. 141 for *Spondylus princeps* (Tucson, University of Arizona Press: 1980).

4. The Pliocene shoreline on Isla Monserrate sits at an elevation of 670 ft (204 m) above sea level, where fossils of *Nerita scabricosta* are common. As described in chapter 8 of *Off-Trail Adventures in Baja California* (Tucson: University of Arizona Press, 2014), 167, few visitors have climbed to the roof of Isla Monserrat to experience a former rocky shore so elegantly delineated by an unconformity between limestone and Miocene andesite.

5. Brusca in *Common Intertidal Invertebrates of the Gulf of California*, 158, suggests from preliminary observations that a gradient occurs in the Gulf of California, whereby those *Nerita scabricosta* living in the south grow to a larger size than those in the north.

6. Detailed estimates for ice volumes during the most recent continental glaciations were applied by W. L. Donn and others from the Lamont Geological Observatory to approximate the amount of sea-level fall worldwide. Find their

1962 article on "Pleistocene Ice Volumes and Sea-Level Lowering," published in the *Journal of Geology* 70:206–14. Such calculations generally agree with the mapping of ancient beach terraces as well as former peat bogs now more than 328 ft (100 m) below sea level.

7.   Paleontological details of the Pleistocene coral reef at Punta Chivato are laid out by M. E. Johnson in chapter 4 (pp. 67–70) of *Discovering the Geology of Baja California* (Tucson: University of Arizona Press, 2002).

8.   Paleontological details of the Pleistocene coral reef on Isla Coronados are described in more detail by M. E. Johnson in chapter 7 (pp. 152–158) of *Off-Trail Adventures in Baja California* (Tucson: University of Arizona press, 2014).

Chapter 10

1.   The toll on road infrastructure as a result of the 1998 Hurricane Mitch is described in the 2002 contribution on "Digital Inventory of Landslides and Related Deposits in Honduras Triggered by Hurricane Mitch," by E. L. Harp, K. W. Hagaman, M. D. Held, and J. P. McKenna, available in *United States Geological Survey Open-File Report* 02–61.

2.   Ed Ricketts's embrace of "non-teleological thinking" is marked by a well-known passage regarding the Asian concept of "being" found in *The Log From the Sea of Cortez*. More on this connection is discussed in the epilogue on Zen aesthetics in *Off-Trail Adventures in Baja California* (Tucson: University of Arizona Press, 2014), 207–10. The consummate fisherman Ray Cannon was also attracted to Buddhism and other Asian philosophies, as discussed in Gene Kira's *The Unforgettable Sea of Cortez* (Torrance, Calif.: Cortez Publications, 1999). See especially p. 9.

3.   Refer to the Annual Mean Precipitation chart for the Baja California peninsula (p. 5) for the Cape Region in the third edition of the *Baja California Plant Field Guide*, by Jon P. Redman and Norman C. Roberts (San Diego: San Diego Natural History Museum and Sunbelt Publications, 2012).

4.   News report from the English-language *Mexican News Daily* under the byline "Torrential Rains Cause Damage in Mulegé," published September 5, 2016.

5.   Although outdated, the most extensive summary available in English on the mainland Chinese system of coastal barriers is reported by C. Li, D. Fan, B. Deng, and V. Korotaev in "The Coasts of China and Issues of Sea Level Rise," *Journal of Coastal Research* 43 (2004): 36–49.

Primary References

Avila, L. 2016. The 2015 Eastern North Pacific Hurricane Season: A Very Active Year. *Weatherwise* 69:36–42.

Backus, D. H., and M. E. Johnson. 2014. Stromatolitic Mats from an Uplifted Pleistocene Lagoon at Punta Chivato on the Gulf of California (Mexico). *Palaios* 29:460–66.

Backus, D. H., M. E. Johnson, and R. Riosmena-Rodrígues. 2012. Distribution, Sediment Source, and Coastal Erosion of Fan-Delta Systems on Isla Cerralvo (Lower Gulf of California, Mexico). *Journal of Coastal Research* 28:210–24.

Bigioggero, B., S. Chiesa, A. Zanchi, A. Montrasio, and L. Vezzoli. 1995. The Cerro Mencenares Volcanic Center, Baja California Sur: Source and Tectonic Control on Post-subduction Magmatism Within the Gulf Rift. *Geological Society of America Bulletin* 107:1108–22.

Boyle, T. C. 2000. *A Friend of the Earth*. New York: Penguin Putnam.

Brierley, C. M., A. V. Fedorov, Z. Liu, T. D. Herbert, K. T. Lawrence, and J. P. LaRiviere. 2009. Greatly Expanded Tropical Warm Pool and Weakened Hadley Circulation in the Early Pliocene. *Science* 323:1714–18.

Brusca, R. C., and D. A. Thomson. 1975. Pulmo Reef: The Only "Coral Reef" in the Gulf of California. *Ciencias Marinas* 2:37–53.

Cheng, L., J. Abrahan, J. Zhyu, K. E. Trenberth, J. Fasullo, T. Boyer, R. Locaarnini, et al. 2020. Record-Setting Ocean Warmth Continued in 2019. *Advances in Atmospheric Sciences* 37:137–42.

DeDiego-Forbis, T., R. Douglas, D. Gorsline, E. N. Nava-Sanchez, L. Mack, and J. Banner, 2004. Late Pleistocene (Late Interglacial) Terrace Deposits, Bahía Coyote, Baja California Sur, Mexico. *Quaternary International* 120: 29–40.

Eckman, T. 2018. Mexican Middle Schoolers Model Their Local Watershed. *In the Trenches: The News Magazine of the National Association of Geoscience Teachers* 8:1–2.

Emhoff, K. F., M. E. Johnson, D. H. Backus, and J. Ledesma-Vázquez. 2012. Pliocene Stratigraphy at Paredones Blancos: Significance of a Massive Crushed-Rhodolith Deposit on Isla Cerralvo, Baja California Sur (Mexico). *Journal of Coastal Research* 28:234–43.

Eros, J. M., M. E. Johnson, and D. H. Backus. 2006. Rocky Shores and Development of the Pliocene-Pleistocene Arroyo Blanco Basin on Isla Carmen in the Gulf of California, Mexico. *Canadian Journal of Earth Sciences* 43:1149–64.

Johnson, E. M., B. G. Baarli, M. E. Johnson, and P. R. Gomez-Zuñiga. 2020. Effects of Hurricanes, Mudslides, Flooding, and Riverine Erosion on the Erasure of Archaeological Sites in Tropical Highland Honduras. *Geoarchaeology* 35:338–350.

Johnson, M. E. 2014. *Off-Trail Adventures in Baja California: Exploring Landscapes and Geology on Gulf Shores and Islands.* Tucson: University of Arizona Press.

Johnson, M. E., D. H. Backus, A. L. Carreño, and J. Ledesma-Vázquez. 2019a. Rhyolite Domes and Subsequent Offlap of Pliocene Carbonates on Volcanic Islets at San Basilio (Baja California Sur, Mexico). *Geosciences* 9:87. https://doi.org/10.3390/geosciences9020087.

Johnson, M. E., D. H. Backus, and J. Ledesma-Vázquez. 2017. Growth of the Ballena Fan Delta on the Gulf of California (Mexico) at the Close of the Pliocene Warm Period. *Facies* 63:14. https://doi.org/10.1007/s10347-017-0495-y.

Johnson, M. E., D. H. Backus, and R. Riosmena-Rogríguez. 2009a. Contribution of Rhodoliths to the Generation of Pliocene-Pleistocene Limestone in the Gulf of California. In *Atlas of Coastal Ecosystems in the Western Gulf of California*, edited by M. E. Johnson and J. Ledesma-Vázquez, 83–94. Tucson: University of Arizona Press.

Johnson, M. E., and R. J. Cuffey. 1997. Bryozoan Nodules Built Around Andesite Clasts from the Upper Pliocene of Baja California: Paleoecological Implications and Closure of the Panama Isthmus. Geological Society of America Special Paper 318:111–117.

Johnson, M. E., R. Guardado-France, E. M. Johnson, and J. Ledesma-Vázquez. 2019b. Geomorphology of a Holocene Hurricane Deposit Eroded from Rhyolite Sea Cliffs on Ensenada Almeja (Baja California Sur, Mexico). *Journal of Marine Science and Engineering* 7:193. https://doi.org/10.3390/jmse7060193.

Johnson, M. E., E. M. Johnson, R. Guardado-France, and J. Ledesma-Vázquez. 2020. Holocene Hurricane Deposits Eroded as Coastal Barriers from Andesite Sea Cliffs at Puerto Escondido (Baja California Sur, Mexico). *Journal of Marine Science and Engineering* 8:75. https://doi.org/10.3390/jmse8020075.

Johnson, M. E., and J. Ledesma-Vázquez. 1999. Biological Zonation on a Rocky-Shore Boulder Deposit: Upper Pleistocene Bahía San Antonio (Baja California Sur, Mexico). *Palaios* 14:569–84.

Johnson, M. E., J. Ledesma-Vázquez, and D. H. Backus. 2016. Tectonic Decapitation of a Pliocene Mega-delta on Isla del Carmen in the Gulf of California (Mexico): And a River Ran Through It. *Journal of Geology* 124:556–74.

Johnson, M. E., J. Ledesma-Vázquez, and R. Guardado-France. 2018. Coastal Geomorphology of a Holocene Hurricane Deposit on a Pleistocene Marine

Terrace from Isla Carmen (Baja California Sur Mexico). *Journal of Marine Science and Engineering* 6:108. https://doi.org/10.3390/jmse6040108.

Johnson, M. E., J. Ledesma-Vázquez, and A. Y. Montiel-Boehringer. 2009b. Growth of Pliocene-Pleistocene Clam Banks (Mollusca, Bivalvia) and Related Tectonic Constraints in the Gulf of California. In *Atlas of Coastal Ecosystems in the Western Gulf of California*, edited by M. E. Johnson and J. Ledesma-Vázquez, 104–16. Tucson: University of Arizona Press.

Johnson, M. E., J. Ledesma-Vázquez, D. H. Backus, and M. R. González. 2012. Lagoon Microbialites on Isla Angel de la Guarda and Associated Peninsular Shores, Gulf of California (Mexico). *Sedimentary Geology* 263/264:76–84.

Johnson, M. E., R. A. López-Pérez, C. R. Ransom, and J. Ledesma-Vázquez. 2007. Late Pleistocene Coral-Reef Development on Isla Coronados, Gulf of California. *Ciencias Marinas* 32:105–20.

Kennedy, A., N. Morik, T. Yasuda, T. Shimozono, T. Tomiczetk, A. Donahue, T. Shimura, and Y. Imai. 2017. Extreme Block and Boulder Transport Along a Cliffed Coastline (Calicoan Island, Philippines) During Super Typhoon Haiyan. *Marine Geology* 385:65–77.

Kira, G. S. 1999. *The Unforgettable Sea of Cortez: Baja California's Golden Age, 1947–1977; The Life and Writings of Ray Cannon*. Torrance, Calif.: Cortez.

Kirkland, D. W., J. P. Bradbury, and W. E. Dean, Jr. 1966. Origin of Carmen Island Salt Deposit Baja California, Mexico. *Journal of Geology* 74:932–38.

Kossin, J. P., K. R. Knapp, T. L. Olander, and C. S. Welden. 2020. Global Increase in Major Tropical Cyclone Exceedance Probability Over the Past Four Decades. Proceedings National Academy of Science (PNAS). www.pnas.org/cgi/doi/10.1073/pnas.1920849117.

Kozlowski, J. A., M. E. Johnson, J. Ledesma-Vázquez, D. Birgel, J. Peckmann, and C. Schleper. 2018. Microbial Diversity of a Closed Salt Lagoon in the Puertecitos Area, Upper Gulf of California. *Ciencias Marinas* 44:71–90.

Krutch, J. W. 1961. *The Forgotten Peninsula: A Naturalist in Baja California*. Tucson: University of Arizona Press.

López-Pérez, R. A., and A. F. Budd. 2009. Coral Diversification in the Gulf of California During the Late Miocene to Pleistocene. In *Atlas of Coastal Ecosystems in the Western Gulf of California*, edited by M. E. Johnson and J. Ledesma-Vázquez, 58–71. Tucson: University of Arizona Press.

Lyle, M., and G. E. Ness. 1991. The Opening of the Southern Gulf of California. In *The Gulf and Peninsular Province of the Californias*, edited by J. P. Dauphin and B. Simoneit, 403–23. American Association of Petroleum Geologists Memoir 47.

Mark, C., S. Gupta, A. Carter, D. F. Mark, C. Gautheron, and A. Martin. 2014. Rift Flank Uplift at the Gulf of California: No Requirement for Asthenospheric Upwelling. *Geology* 42:259–62.

McFall, C. C. 1968. *Reconnaissance Geology of the Concepcion Bay Area, Baja California, Mexico*. Stanford: Stanford University Publications, Geological Sciences.

Merril, R. J., and E. S. Hobson. 1970. Field Observations of *Dendraster excentricus*, a Sand Dollar of Western North America. *American Midland Naturalist* 83:595–624.

Muriá-Vila, D., M. A. Jaimes, A. Pozos-Estrada, A. López, E. Reinoso, M. M. Chávez, F. Peña, J. Sánchez-Sesma, and O. López. 2018. Effects of Hurricane Odile on the Infrastructure of Baja California Sur, Mexico. *Natural Hazards* 9:963–81.

Oliver, E. C. J., M. G. Donat, M. T. Burrows, P. J. More, D. A. Smale, L. V. Alexander, J. A. Benthuysen, et al. 2018. Longer and More Frequent Marine Heatwaves Over the Century. *Nature Communications* 9:1324.

Reyes-Bonilla, H., and R. A. López-Pérez. 2009. Corals and Coral-Reef Communities in the Gulf of California. In *Atlas of Coastal Ecosystems in the Western Gulf of California*, edited by M. E. Johnson and J. Ledesma-Vázquez, 45–57. Tucson: University of Arizona Press.

Sewell, A. A., M. E. Johnson, D. H. Backus, and J. Ledesma-Vázquez. 2007. Rhodolith Detritus Impounded by a Coastal Dune on Isla Coronados, Gulf of California. *Ciencias Marinas* 33:483–94.

Shepard, F. P. 1950. Part 3: Submarine Topography of the Gulf of California. In *1940 E. W. Scripps Cruise to the Gulf of California*. Geological Society of American Memoir 43. N.p.: Geological Society of America.

Squires, D. F. 1959. Results of the Puritan-American Museum of Natural History Expedition to Western Mexico. Part 7: Corals and Coral Reefs in the Gulf of California. *Bulletin American Museum of Natural History* 118:367–432.

Steinbeck, J., and E. F. Ricketts. 1941. *Sea of Cortez: A Leisurely Journal of Travel and Research*. New York: Viking Press.

Stock, J. M. 2000. Relation of the Puertecitos Volcanic Province, Baja California, Mexico, to Development of the Plate Boundary in the Gulf of California. *Geological Society of America Special Paper* 334:143–56.

Tierney, P. W., and M. E. Johnson. 2012. Stabilization Role of Crustose Coralline Algae During Late Pleistocene Reef Development on Isla Cerralvo, Baja California Sur (Mexico). *Journal of Coastal Research* 28:244–54.

Umhoefer, P. J., R. J. Dorsey, S. W. Willsey, L. Mayer, and P. Renne. 2001. Stratigraphy and Geochronology of the Comondú Group near Loreto, Baja California Sur, Mexico. *Sedimentary Geology* 1445:125–47.

Wallace-Wells, D. 2019. *The Uninhabitable Earth: Life After Warming*. New York: Tim Duggan Books.

Wara, M. W., A. C. Ravelo, and M. L. Delaney. 2005. Permanent El Niño Conditions During the Pliocene Warm Period. Science. 309:758–61.

Zwinger, A. 1983. *A Desert Country near the Sea*. Tucson: University of Arizona Press.

Index

About the Author

Markes E. Johnson is the Charles L. MacMillan Professor of Natural Science, Emeritus, at Williams College in Williamstown, Massachusetts, where he taught courses in historical geology, paleontology, and stratigraphy in the Geosciences Department over a thirty-five-year career. He grew up in the Midwest, where dolomite bluffs along the Upper Mississippi River drew his attention as a hobbyist, collecting marine fossils that lived in a vast continental sea. His undergraduate education in geology concluded with a BA degree (1971) from the University of Iowa. His advanced training in paleoecology through the Department of Geophysical Sciences at the University of Chicago culminated with a PhD degree (1977). Since 1989, Professor Johnson has made one or two trips a year to the Baja California peninsula and Mexico's Gulf of California to study coastal deposits related to the Pliocene Warm Period and later Pleistocene Epochs when sea level and global temperatures were higher than today. Since 2009, he has been active with studies regarding the Miocene to Pleistocene history of many North Atlantic islands, including those of the Cape Verde, Canary, Madeira, and Azores archipelagos. Professor Johnson was the recipient of a 2011 Nelson Bushnell Prize for excellence in scholarship and teaching at Williams College. He is a frequent guest lecturer, accompanying excursions sponsored by the Williams College Society of Alumni to Scandinavia, the Iberian Peninsula, the Galapagos Islands, and Mexico's Gulf of California.